ORAL COMMUNICATION

IN TECHNICAL PROFESSIONS AND BUSINESSES

Diane O. Casagrande

West Chester University

Roger D. Casagrande

General Electric Company

Wadsworth Publishing Company, Belmont, California/A Division of Wadsworth, Inc.

For Bradley, Jeffrey and Kimberly . . .

three other of our cooperative projects

Communications Editor *Kristine Clerkin*
Editorial Assistant *Naomi Brown*
Production Editor *Gary Mcdonald*
Managing Designers *Cynthia Bassett, Merle Sanderson*
Print Buyer *Karen Hunt*
Designer *Jim Gray*
Copy Editor *Joan C. Pendleton*
Technical Illustrators *Jeffrey Casagrande, Ginny Mickelson*
Cover *Salinda Tyson*

Printed in the United States of America
1 2 3 4 5 6 7 8 9 10---90 89 88 87 86

ISBN 0-534-05532-X

Library of Congress Cataloging-in-Publication Data

Casagrande, Diane O.
Oral communication in technical professions and businesses.

Includes bibliographies and index.
1. Communication of technical information. 2. Communication in management.
3. Oral communication.
I. Casagrande, Roger D. II. Title.
T10.5.C37 1986 601'.41 85-16841
ISBN 0-534-05532-X

CONTENTS

Preface x

PART I
FUNDAMENTALS FOR TECHNICAL COMMUNICATION 1

Chapter 1
Communication for the Technical Professional **3**

Communication As Part of Life 6
Random Sound Perceptions 6 ▪ Self-Talk 7 ▪ One-to-One Encounters 8 ▪ Group Encounters 9 ▪ Public Encounters 9
Communication To Meet Professional Expectations 11
Expediency 11 ▪ Credibility 12 ▪ Success 14
Activities 14
Notes 15

Chapter 2
Communication Models **17**

The Constant Variables in Communication 19
The Sender 20 ▪ Background 21 ▪ Stimuli: Random and Ordered 22 ▪ Channel 23 ▪ The Receiver 24 ▪ Feedback 26 ▪ Entropic Noise 29 ▪ Setting 31 ▪ The Model of Communication 32
The Transient Variables in Communication 34
Experiential Effect 34 ▪ Perceptual Bias 35 ▪ Professional Expectations 38
The Communication Process Model 40
Wrap-up 42
Activities 43
Notes 44

Chapter 3
The Symbol Systems of Communication 45

Random and Ordered Stimuli: Signs and Symbols 45
Symbols and Language 47
Nonverbal Communication 48
Paralanguage 50 ▪ Objects and Color 52 ▪ Touch 55 ▪ Eye Contact 56 ▪ Proxemics 58 ▪ Kinesics 60 ▪ Cautions 61
Verbal Communication 62
Inevitability of Change 63 ▪ Levels of Abstraction 65 ▪ Fact/Inference 69
Wrap-up 75
Activities 76
Notes 77

Chapter 4
Listening 79

Listening to Yourself: Internal Feedback 82
Listening to Others: External Feedback 83
Listening Effectively 84
Nine Techniques for Good Listening 84 ▪ Two Checks for Good Listening 90
Wrap-up 91
Activities 92
Notes 93

Summary: Part I
Fundamentals for Technical Communication 94

PART II
PREPARATION FOR ENCOUNTER 97

Chapter 5
Intrapersonal Communication 99

The Versions of Me 100
My Version of Me 102 ▪ My Version of the Me You Know 102 ▪ The Hidden Version 102 ▪ Your Version of Me 102 ▪ Your Version of My Version of Me 103
Human Values and Needs 104
Values 104 ▪ Needs 106
Dealing with Anxiety 109
Speech Anxiety 109 ▪ Fear 110 ▪ Quietness and Shyness 113
Wrap-up 117
Activities 117
Notes 118

Chapter 6
From Self to Others: Audiences 120
Analyzing Audiences 121
Demographics 122 ▪ Attitudes 123
Receiver Response and Speaker Goals 128
Traditional Speaker Goals 129 ▪ Contemporary Speaker Goals 130
Goals in the Professional World 134
Acceptance of Image 134 ▪ Acceptance of Ideas 136 ▪ Acceptance of Methodology 138 ▪ Acceptance of Contract 139
Wrap-up 141
Activities 142
Notes 143

Chapter 7
Research and Organization of Presentations 144
Audience Analysis and Topic Selection 145
Mandated Topics 145 ▪ Personal-Choice Topics 146
Topic Research 147
Written Materials 147 ▪ Interview Materials 149 ▪ Personal Experiences 150 ▪ Consultants 150
Topic Organization 151
Beginnings and Endings 152 ▪ Body of the Speech 153 ▪ Patterns, Purposes, and Audiences 158
Wrap-up 158
Activities 159
Notes 160

Chapter 8
Proving: Evidence and Reasoning 161

Components of Proof 163
Sources of Proof 164
Ethos 164 ▪ Pathos 164 ▪ Logos 165
Evidence: A Documentation Function 165
Facts 165 ▪ Opinions 166 ▪ Strong and Weak Documentation 167
Reasoning: A Linking Function 170
Strong Links: Categories and Types of Reasoning 171
Weak Links: Fallacies 175
Testing for Strength of Linkage 178
Wrap-up 181
Activities 182
Notes 183

Chapter 9
Communication Aids 184

Aids in Technical Communication 184
Guidelines for Effective Aids 185
Make It Clear 186 ▪ Make It Interesting 186 ▪ Exaggerate Reality 187 ▪ Think in Visual Terms 187 ▪ Prepare the Audience 188 ▪ Keep Illustrations Uncluttered 189 ▪ Involve the Audience 189
Types of Aids for Technical Presentations 190
Presentational Charts 190 ▪ Objects and Models 201 ▪ Paper/Flip Charts 202 ▪ Photographs and Films 202
Communication Equipment in Modern Corporations 204
Telecommunications 204 ▪ Remote Computer Terminals 205 ▪ Microcomputers 205
Wrap-up 206
Activities 206
Notes 207

Summary: Part II
Preparation for Encounter 208

PART III
SETTINGS FOR TECHNICAL COMMUNICATION 211

Chapter 10
One-to-One: Dyads 213

Interviewing: The Initial Dyad 213
Interview Preparation 214 ▪ Questions and Answers 216 ▪ Interview Courtesies 218
Career Development: Maintenance Dyads 219
Interpersonal Relations 219 ▪ Networks 220 ▪ Conflict Management 221
Power Structures: Give-and-Take Dyads 226
Power and Superior-Subordinate Interaction 227 ▪ Power and Giving or Taking Orders 232
Wrap-up 235
Activities 235
Notes 236

Chapter 11
Small-Group Theory 238

Components of Groups 239
Numbers 239 ▪ Purpose 240 ▪ Norms 240 ▪ Cohesion 240 ▪ Consensus 241 ▪ Special Collections of People 241
Types of Groups 242
Brainstorming 242 ▪ Spontaneous Meetings 242 ▪ Panels and Symposia 243 ▪ Forum 243 ▪ Task Meetings 243 ▪ Agenda Problem-Solving Meetings 244
Disadvantages and Advantages of Group Process 244
Disadvantages 244 ▪ Advantages 246
Types of Discussion Questions 248
Questions of Fact 249 ▪ Questions of Value 249 ▪ Questions of Policy 250
Corporate Forms of Group Meetings 251
In-House Meetings 251 ▪ Intercompany Meetings 258 ▪ International Variations 260

Wrap-up 263
Activities 264
Notes 265

Chapter 12
Small-Group Participation and Leadership 266

Formal Meetings and Formal Preparation 266
Meeting Agenda 266 ▪ *Types of Agendas* 268 ▪ *Dewey-Based Problem-Solving Agenda* 268 ▪ *Parliamentary Procedure* 272
Leadership and Participation 279
Roles in Groups 279 ▪ *Leadership in Groups* 281
Participation and Responsibilities 289
Cooperation and Consensus 289 ▪ *Negotiation* 290 ▪ *Preparation* 291
Wrap-up 291
Activities 292
Notes 293

Chapter 13
Public Communication 294

Focused Variables 296
Formalized Goal 296 ▪ *The Specific Audience* 297 ▪ *The Nonverbal Messages* 297 ▪ *The Verbal Messages* 298 ▪ *Choices of Delivery* 299 ▪ *Visual and Audio Aids* 302
Addressing the Professional Audience 302
Professional Organizations 303 ▪ *Sales/Contract Opportunities* 305
Addressing the General Public Audience 308
Addressing the Mass Media Audience 310
Television 310
Wrap-up 313
Activities 315
Notes 316

Chapter 14
Expert Testimony 318

Experts and Sources of Competence 319
Education 320 ▪ *Experience* 320 ▪ *Physical Position* 321 ▪ *Testing for Expertise* 321

Experts in Law Courts 324
Assumptions about Testimony 324 ▪ *Preparation for Testimony 328*
Experts in Hearings 332
Similarities to Law Courts 334 ▪ *Admission of Biases 334*
Wrap-up 336
Activities 337
Notes 339

Summary: Part III
Settings for Technical Communication 340

Glossary 342
Index 351

PREFACE

The beginning of this book goes back further than many of its readers' lifetimes. It began when we, your authors, Diane and Roger, registered as freshmen at Wayne State University in Detroit. Roger was an engineering student and, quite coincidentally, the Engineering Building was located across the street from the Speech Communication office. Diane declared a speech communication major and spent a lot of time at that office, especially after noticing the intriguing young man across the way in the physics lab. As these things happen, we met, argued philosophy and physics, married, and celebrated our twenty-fifth wedding anniversary about midway through the writing of this book.

Between the lab and the anniversary, Roger received his doctorate from Rensselaer Polytechnic Institute and joined General Electric, where he is now a program manager with Spacecraft Operations. Diane's doctorate is from Temple University, and she is professor of Speech Communication at West Chester University. Neither of us has ever stopped discussing philosophy or physics, and each of us has affected the career of the other. Roger taught Diane to be conversant in the jargon and orientation of the technical world. Diane convinced Roger that communication is the focus of all human activity, including science. We both recognized that, as Roger's responsibilities grew in a technical corporation, he found himself increasingly on public platforms, at hearings and proposal presentations. The need for skill in effective oral communication became a priority.

Watching Roger's growth and attending many technical conferences with him, Diane became convinced that the usual college oral-communication course missed the needs of science, engineering, and computer/mathematics students. Furthermore, she realized that *most* people who do not have technical backgrounds eventually work with those who have them. Neither the technical professionals nor the "others" seem to understand one another very well. As a result, Diane went to the administrators of those departments on her campus that prepare technical students and asked them to describe what their majors do after graduation. She especially wanted to find what kind of oral communication would help make postgraduate activities successful when technical and nontechnical people work together. Roger did his research in corporations and compared the academicians' comments with those of his sources. Our joint findings confirmed Diane's suspicion: the rarefied atmosphere of modern technical corporations requires specialized communication knowledge.

Diane designed an experimental section of the Speech Fundamentals course called Speech Communication for the Technical Professional. It is now required of all physics students at West Chester who are preparing to be engineers. When she looked for an appropriate book, she found no current oral communication text written for and to the technical professional. There are many good *writing* texts for such people, but oral communication has been neglected. That seemed peculiar, since study after study documents the need for competent oral communication, especially among scientists/mathematicians/computer specialists. Excellent job opportunities in these disciplines are not lost on students, for they have shown renewed interest in the "hard sciences," engineering programs, and computer/mathematics majors. The need for a book on oral communication written by partners who understand speech communication and the realities of the technical world culminated in the text you are now reading.

We have written our book for undergraduate technical students and those who foresee themselves working with technical people. We assume that any science major and most nonscience majors must understand the demands of modern technical communication, for they will all be subject to those demands. Our goal is to balance theory and practical corporate experience. "Purely practical" books tend to become "How to" or formula approaches, and we just do not believe human communication lends itself to prescription formulas. "Highly theoretical" books, on the other hand, seldom allow the student to recognize how the theory translates directly into real experiences.

We hope we have achieved this balance, for all the examples from the corporate world are real ones, many taken from direct observation. Because Roger knows General Electric best, ideas grew from his experience there, but we found parallels widespread in both our research and the literature. All parts of the book have been tested in West Chester classrooms. In many ways, the students who reacted to one or another of the topics serve as "auxiliary authors."

The topics we include are comprehensive, purposefully so. Each is essential to technical professionals moving up through their corporations. No single semester is long enough to cover everything students need. We think that might be fine too. Instructors can choose among the topics to tailor very different courses than their colleagues might. If we have done our job, students should be convinced that communication really is a tool for success. This tool can be refined if students keep the book until they need some communication skill and then turn to the book for guidance.

We recognize that the topics we cover and our approach to them are different from those of other books. Our selections are based on the best advice and experience of persons working in various technical disciplines, in both the United States and Europe. As a result, some topics appear for the first time in an introductory speech book. For example:

- The types of corporate technical meetings
- International adaptations to group discussion
- The pervasiveness of public communication
- Expert-witness preparation
- The hybrid public speech that corporations develop
- Proving as a persuasive requirement in science
- An original model of self

Some topics have been adapted for the special use of technical people. Among them:

- Parliamentary procedure as a frequently used contract-discussion requirement
- Science orientation and language
- Visual aids technical people use
- Interviewing for technical positions
- An original communication model using concepts familiar to those trained in science

In each case, we have asked, "How much of this does the technical professional actually need?" If the information was needed on the job or as part of normal human functioning off the job, we included it. Our focus is the training of a prospective scientist/mathematician/computer specialist for success in the corporate world and satisfaction beyond the business day. Our premise is that communication competence is the key to achieving that success and satisfaction. We think it is now time for students to make this discovery. We offer the uniqueness of our long-standing partnership in speech communication and technical corporate experience to guide students. We hope they find the following chapters challenging, eye-opening, and useful. If there is a special someone across the way with whom they would like to argue some of the points, we think that would be fine too.

ACKNOWLEDGMENTS

The British dramatist-poet Colley Cibber wrote, "Words are but empty thanks." He was right, of course, but words are the only way we have to give recognition to those whose help has been so vital. We do not mean such words to be empty, but we certainly recognize that they are inadequate.

To Kristine Clerkin, our editor, and the excellent staff at Wadsworth, especially Gary Mcdonald, and copy editor Joan Pendleton, our debt is most profound. It was Kris who first saw the possibilities for a book by the partnership of unlikely disciplines we represent. She, Gary and Joan brought the idea to reality. Kris found the reviewers, all of whom left the imprint of their suggestions on every page: Richard H. Arthur, Montana State University; Janet Engelbert, Heald College, San Francisco Business Division; Susan Brown Fiechtner, Texas A & M University; Jan Kelly, San Jose State University; Stephen King, San Diego State University; Michael R. Moore, Purdue University–Calumet; Roberta K. Ray, Montana College of Mineral Science and Technology; Grant Savage, Texas Tech University; and Tom Wirkus, University of Wisconsin–La Crosse.

David Oaks and Michael Dooley gave us the wealth of their editorial experience at every step of preparation. How they found time in their own lives to help us so much is mystifying. That they did so proves the strength of sibling relationships. To our biologic brother, Dave, and adopted brother, Mike, thanks beyond words.

We make special note with parental pride of the use we made of our children during the preparation of this manuscript. Most of the model drawings were prepared by our son Jeffrey. Jeff is a newly graduated electrical engineer, and we tried out topics and approaches on him throughout. The communication model is as much his work as it is ours, and we acknowledge his skill with professional pride. Our daughter, Kimberly, prepared the drawing in Chapter Three. It was also she who had the patience to go through repeated proofreading when the rest of us had given up. Our son Bradley and daughter-in-law Deborah encouraged us when our lack of availability must have been frustrating. To all our children, we owe a debt of love.

Scott Anderson and Kate O'Shea, former West Chester University students, served as our photographers. These talented people took many pictures to show visually what we want to say orally. We decided not to illustrate most of the chapters, but Scott and Kate have our admiration.

A debt of another kind extends to many people who recognized our limited time and understood what that meant to missed engagements and obligations. To co-workers at General Electric and West Chester University, our deepest gratitude. For Ed and Alice Oaks, especially, thank you for being there.

As we feared, Cibber was right and the words are inadequate. The feelings they attempt to describe, however, are real.

DOC
RDC
Malvern, Pennsylvania

PART I

FUNDAMENTALS FOR TECHNICAL COMMUNICATION

CHAPTER ONE

COMMUNICATION FOR THE TECHNICAL PROFESSIONAL

Fantasizing about a career is one of many exciting and often frustrating college experiences. Your interest in some course of study often leads to speculation about a career in that discipline. Your mental image and the reality might be very different. Alice, the biology student, for example, who pictures herself in a quiet laboratory surrounded by up-to-date equipment, might be stunned to observe the methods and working conditions of a modern lab. Perhaps, through a field study or an internship, she discovers that her picture of a research biologist and the actuality of the work have little similarity. At that point, she has to make some serious decisions. Is the real example satisfactory enough for her to change her mental image, or is the actual so disappointing that a new career has to be found?

One goal of this book is to help students visualize a technical career as it develops in modern corporations. Those who plan such careers and those who will hire them and work with them must learn to understand each other if the career and the business are to be successful. We believe that communication is the heart of such understanding. We the authors and you the reader will be looking at two themes that support the structure of these chapters:

- Most technical professionals function in corporate settings, some large, some small. These technical specialists and the people who work with them in national or international industries must understand each other's orientations to work together harmoniously.
- Technical professionals live in a communicative world, and their competence in communication helps or hinders their work and their dealings with coworkers.

The temptation to separate people into categories of *technical* and *others* is persistent. In colleges the separation widens. Technical majors such as the "hard sciences," engineering, mathematics, and computer sciences are filled with content of enormous complexity. Usually little time remains for "other" courses beyond the core of general education requirements. With training in physics and calculus or thermodynamics and differential equations comes more than subject matter. This training also orients students to look at the world in a particular way, which sometimes separates them from *others*. Some of that separation is simply a matter of logistics. Chemistry, after all, must be taught in places where chemical equipment is available. After the introductory course, the only persons who tend to see each other in the buildings where the equipment is present are the other chemistry majors. Somewhere along the line, the combination of the orientation and the physical distance brings the perception that those *others* really *are* different.

Perhaps it is the predictable result of concentrating on some area of study, technical or not. The whole world begins to revolve around that study. You declare a physics major and, seemingly by magic, you see physical processes all around. Computer scientists claim that computer data "interfaces with everything." Prospective chemists just know that their science represents the truly important. For many centuries, mathematicians have called their discipline the "Queen of the sciences." Perhaps this bias is all as it should be; shouldn't you care enough about your chosen area of study to be protective of it and excited by its possibilities?

To some degree, all students identify with their major field of study. For the technical student, when the problem of logistics and the increasing constraint of the orientation is added, a restrictive attitude seems to surface: *any* other subject matter becomes unimportant and irrelevant. "What's that got to do with engineering?" "What's more important than the *truth* of science?" "If I could take the six credits I have to use in history and take two more math classes instead, that would really help me."

While you are in college, such perceptions might not cause too much difficulty. Plenty of chemistry majors as well as physics and computer-science students speak a similar language. The problem begins when the student becomes a technical professional and moves into the corporate world. The very training that made Ed, the engineering student, a fine engineer might now cause him trouble as he deals with the marketing staff, the financial department, and those in public relations. They in turn conclude, "Don't talk to Ed the Engineer. Nobody can figure out what he means."

Ed or his counterpart Alice, the biologist, or Jason, the computer specialist, or Sarah, the astronomer, all might be competent subject-content specialists. Their problem and the problem of those who must work with them is *communication.* The exchange of ideas, feelings, and events is the fundamental behavior of human beings. Our goal, in this book, is to help you understand the process of communication and how it can be improved in the corporate/social setting.

We will leave to others the arguments about what one needs in a science curriculum to be an educated person. The task ahead, for both you and the two of us writing this book, is to think about communication in the world of technology. We invite you to read our preface, which is our means of introducing ourselves to you. Both of us are convinced that we know the world of technology as it operates in the corporate structure. Our bias, you should know at the outset, is obvious: communication competence is at the heart of a successful career. You, however, may need to be convinced of that.

In the mid-1920s, a researcher named Paul Rankin published the rather startling results from one of his studies. He determined the percentage of time people spent on the four communication activities then recognized as being performed in a literate society:[1]

Activity	*Percentage of time*
Listening	42.1
Speaking	31.9
Reading	15.0
Writing	11.0

Since then, *nonverbals* have come to be recognized as another communicative activity. We note this addition, but Rankin's results still provide cause for serious thought. As Wolvin and Coakley summarize in their book, *Listening*, at least three recent investigations have verified Rankin's findings, using the same four communication activities.[2] That data may seem inflated until you monitor yourself hour by hour. Even in the college years, when you tend to do concentrated reading and writing, how does your day pass? Consider: you spend many hours listening to your professors (who are, for the most part, talking). You mumble to yourself as you sit figuring out a homework assignment. You listen to your roommates, friends, and customers on the job. You spend a long time alone listening to music or random sound, all the while engaged in a very active talk-think conversation with yourself. Go ahead; test the data scientifically observing your own day. You will find, as did most contemporary investigators who duplicated and improved on Rankin's experimental design, that the numbers are verified.

Most of us are unaware of the constant listening/speaking conversations we have with ourselves and others. You sit in classes and think you have been writing constantly for fifty minutes or reading without stop for hours. You forget that you are listening to *someone* who is talking or to sounds that are flowing around you. While we read, our minds are alive with dialogue. Go ahead; check out the formula.

Rankin's claim may have been startling nearly sixty years ago; but, in comparison to modern thinking about communication, Rankin was a conservative underestimator! Human beings, recent theory asserts, are in constant communication with themselves and their environment. Sleep does not turn the communication off and neither does unconsciousness. Growing knowledge about such things as the intricacy of biological messages sent to organs, the nature of conscious and subconscious thought patterns, and the sending of messages through nonverbal behaviors all result in the most comprehensive view of communication possible: *You cannot not communicate.* That concept will be the focus of the following chapters, and we plan to trace it through the professional world of science and mathematics. Our premise is that, if all human beings are engaged in constant communication, the technical professional must be aware of how this ever-present activity affects the scientific and corporate communities and is, in turn, affected by it. Consider the various communication patterns that constitute daily life.

COMMUNICATION AS PART OF LIFE

Random Sound Perceptions

Not all communication involves humans. Birds signal their presence and challenge to other birds. Lurking cats might become chance receivers of messages intended for the stranger cardinal who has wandered mistakenly into the territory of a barn swallow. Perhaps the cardinal and the swallow will, in turn, become intended or accidental receivers of the "cat danger" warning of a passing sparrow.

We do not know, nor do animal psychologists agree, if cats and birds engage in rule-structured language behavior. We prefer to consider their cries and body postures as random behaviors that communicate.

Humans, like birds and other creatures, pass through their days surrounded by sights and sounds. Some of these stimuli are intended for us and regulate our behavior: the flashing traffic light and ambulance siren are examples of sights and sounds that alter the actions of those who see and hear them.

Some things we see and hear are not meant for us at all. We could have been the observer of the exchange between the cardinal and the barn swallow instead of the cat; we could have overheard part of a conversation between a couple going to class that was meant for no one but those two people. Sometimes such overheard conversations even cause a change in our own future action. Have you ever overheard someone talking about a movie that you had not intended to see and, based on what you were hearing, decided to see the movie? That was a case of communication intended for someone else which resulted in an alteration of your own behavior. Then, too, we have all been the receivers of music-sound intended for someone else's pleasure, when we wanted quiet for study or sleep. Like random numbers that have no predictable order, random sights and sounds communicate unpredictably.

Even when we are alone, the sights and sounds continue. Electric appliances turn on and off; loose windows send off an irritating clatter; neon-flashing lights attract attention; a playful dog chasing a Frisbee distracts us; the wind makes an interesting moving mobile of the tree branches outside our window. Try listening when you think there is no sound in the room or try watching when you think there is nothing to see. You will discover that communicative stimuli are around you always.

Self-Talk

An auditor is always present when you ask questions or make comments. That auditor is there even if the statements are not made out loud. The witness is you.

Wendell Johnson coined a graphic term to describe your own presence in all your communications. He called it "your most enchanted listener."[3] Although Johnson recognized the *listener* function each of us performs for ourselves, modern theorists argue that we also *send* messages to ourselves, creating an inner conversation. Such dialogues are usually called *intrapersonal communication*, and they are not without potential trouble. Each of us is capable of talking ourselves into extreme behavior—panic, for example. We argue with ourselves; we try out conversations before we voice them aloud; we talk about many things in the constant stream of consciousness we call thought. These "conversations in our heads" are fascinating to psychologists and behaviorists, yet they pale in significance when we realize that our various organic functions are linked to each other by a complex set of messages sent by the brain. In effect, the human body is a physical communicative system in which each part sends messages to and receives messages from the other parts. The concept of biofeedback is based on the theory that a person can send (and learn to monitor) messages to and/or from organic systems previously considered beyond conscious control. Thus, high-blood-pressure patients are taught to recognize the signs that rising blood-pressure levels send and then *talk to their circulatory systems* to keep the levels out of danger. It sounds fanciful or even amusing, but biofeedback has come to be a useful and widely accepted medical process. The message certainly is, "Some communication system is always happening inside and around you."

One-to-One Encounters

Most of us think of communication as happening with someone outside of ourselves. We hope that view has now been altered by the preceding discussion. Assuming you keep intrapersonal communication in mind, it is helpful to gain sensitivity to the *interpersonal* nature of the process. Communication becomes interpersonal when it involves another individual. Whether we call it interpersonal, dyadic, or one-on-one, we are talking about an extremely important chunk of time each person spends each day interacting with one other person.

Many of the popular articles published in widely circulated books and magazines are about one-on-one problems. A quick survey of such articles as "Betrayal: Between Friends," "How to Flirt," "How to Seduce a Man," "Kids' Coded Messages," or "How to Stop Fighting and Start Loving Again,"[4] reveal that many of us are concerned about our intimate or business dyads. That concern is well advised. Our most important personal and business relationships are established and maintained via our communication with that "significant other."

Some one-on-one communications are not so personally important as others, of course. When you stop someone in the city to ask directions or talk about last night's TV show with the redheaded student in your math class or engage in idle chitchat with a coworker during the coffee break, such random encounters do not have the impact an intense conversation with a loved one might have. Nevertheless, they are a necessary and potentially pleasurable part of everyday life. They are also sources of irritation and bad relationships. Much of the reputation each of us has among our friends and acquaintances is the summary of our individual communication encounters with a series of individual persons. It is a part of social and professional life none of us can discount.

Group Encounters

How many persons make a small group or when a small group becomes large are ambiguous questions. For our purposes, *group* refers to your communication patterns with fewer than ten or twelve people when each person has equal opportunity and expectation of speaking. A more formal definition will be suggested in Chapter 11. Much of the problem solving done on many campuses and in business settings is done in groups. For most of us, then, much of our daily time is spent in regularly scheduled meetings or with casually assembled groups.

As we begin to speculate on what happens when several people compete for a finite amount of meeting time, we realize how complex and potentially problem-filled this communication pattern might be.

Public Encounters

When you say "speech" or "communication" to people, chances are they think you mean *public* speech. Majors in speech communication get used to blank stares when they tell their friends that they have no intention of concentrating on the study of *platform speaking*. We hope you will no longer find their statement surprising. Persons communicate, usually orally, in all the settings we have talked about in this chapter, and those settings go far beyond the public platform. For most people, though, speech communication means the stereotypic shaky-kneed speaker addressing fifty people who must remain fairly quiet. Such a setting is *one kind* of speech communication and, as such, warrants our attention, but it is not the only kind.

In-Person Public Encounters People sometimes do give public presentations. Many even seek out the opportunity. Before you decide that you will never do such things, consider the number of presentations any professional discipline requires. Meetings have to be addressed, and new processes must be explained. Some of these meetings are public in the sense that one person has the primary duty of speaking while the remainder listen in. Usually these are settings where both the speaker and the listeners are physically in the same place; such public presentations should be distinguished from those that occur via the mass media.

Public communication does not require large numbers of people and formal environments. The essence of in-person public communication is that one person at a time bears the presentational responsibility for some extended period, while more than two others in the same location serve as message receivers. In this sense, *public communications* are much more common than you might think. The professional world is filled with assignments that require technical specialists to explain their research progress to a monthly meeting of the project team or to act frequently as "expert presenters" of some concept on which a governmental agency panel has requested information. Then, too, all professionals must eventually defend their planning and budget requirements for upper management. Because most of us have assumed that the public speech involves only a stage, a hushed auditorium, and an eloquent speaker, we have not recognized the times in each professional life when all the variables of the public speech come into being. In overlooking these times, we have ignored the abundant research concerning the public platform setting and have not availed ourselves of advice that could help us through our presentations.

Sometimes public communications do conform to stage and audience stereotypes. You might accept the invitation of your school district to explain minicomputers in the classroom to a PTA meeting. You might also be asked by your company to present a slide talk on the applications of domestic solar energy use as part of your company's speakers' bureau. Many of you will probably attempt to get a paper accepted for presentation at the yearly conference of your professional society. All of these are in-person public presentations, and all require your attention and understanding.

Mass-Media Public Encounters Some public presentations take place when the speaker and the audience are physically separated, and a medium such as TV, radio, film, telephone conferences, books, journals, or newspapers bring them together. Although we are not dealing with written modes of mass communication in this book, you are still well-advised to keep them in mind. Radio, TV, and film are very much a part of professional life, as are conferences by telephone. Newscasters often search for "commentary by the professional" to lend credence to their stories. Documentaries are taped routinely in an effort to explain something new or interesting in the research domain. Executive interviews, when a company officer is asked to justify company policy, are far from rare. Each of these is a public mass media presentation. The more ambitious and successful you are, the more you will need to demonstrate competence when that microphone and camera face you.

COMMUNICATION TO MEET PROFESSIONAL EXPECTATIONS

Expediency

For the most part, communication in the technical world is aimed at integrating information into an existing framework so that the new information is both accepted and useful. If you were to consider the development of a new material, for example, and were to describe its characteristics to others, the communication would involve convincing the receivers that the material was within certain accepted parameters. You would have to talk about material strengths under load conditions in different orientations, the thermal parameters of the material, the electrical parameters, chemical stability, and definition of the atomic structure. The basis for including that data as a description of the material is that the professional world uses those parameters for comparing and selecting one option rather than some other.

In the case of the new material, the presentation of the data serves as a very effective means of communication as long as the presenter conforms to the expectation of a discipline. The progress of technology is a story of how the rules have been set for the acceptance of some theory.

Consider another example in the business world: the communication of business objectives. The health of a business is usually defined by a set of financial parameters routinely used by business managers. If you are to talk about business in the professional world, you must be aware of the listener's expectations. To omit data defining expenditures, income, expenses, net profit or loss would make the whole presentation appear incompetent. You need to recognize the expectations of your discipline before you can construct a communication that is understood and accepted. For a business community, the terminology, language, and understanding all reflect the expectations of that audience.

Physician Barbara Gastel points out that *oral* communication for the scientist has at least three overall advantages: the human interest of the scientist as person; the use of hearing, sight, and other senses in presentations; and, unlike writing, immediate two-way interchange. She concludes that oral presentations, "can be among the most effective and satisfying routes for presenting science to the public."[5]

Many technical professionals misunderstand the persuasive nature of research and audience acceptance of published facts. At heart, all human endeavors must be understood and accepted by the audience to whom they are directed. Scientific knowledge is no exception. All past discoveries had to be articulated, presented in the fashion that the current scientific community expected, and finally accepted by the members of those communities. In short, they had to be communicated expediently.

Credibility

One advantage to developing your communication skills in the technical environment is to promote acceptance by your peers and superiors. For the young professional just beginning a career, technical training is a "must," but communication skills might be the determining factor in whether a job will be offered in the first place. Ed Stranix, director of the Philadelphia Computer Learning Center, addresses the question of what makes the ideal candidate for data processing and information systems personnel. He concludes that "the strong candidate should have strength in communication skills, organizational ability, and interpersonal relationships."[6]

In a recent article, columnist Darrell Sifford told the story of a Duke University medical school rejection.[7] The candidate, a superior academic performer with a 4.2 average on a 4.0 scale, could not understand why he had not been accepted. Duke's admissions' personnel did. The young man was a loner and had great difficulty communicating with the outside world: "All he did was study. . . . He had no people skills at all."

Communication skills account for the difference between those who talk and those who talk and are heeded. The contrast between the effectiveness of two people with the same intelligence but different skills in communication can be as striking in the business world as it seems to be in Duke's medical school example. In a study conducted nearly twenty years ago and published in the *Journal of Engineering Education*, experienced engineers were polled about the strengths and shortcomings of their training. Out of 123 areas where additional training should have been given, speech communication ranked fourth. As their experience in the business world increased, the engineers tended to recognize how much communicative skills affected their careers.[8] Gerald Phillips summarized the same point for modern corporations: "Communication skills are so important that they are the most frequent concern of executives looking for talented people to advance in the company."[9]

The recognition of needed communication skill is certainly not new. In 1914 Lillian M. Gilbreth set forth some of the skills that managers in technical disciplines needed in addition to subject-matter competence. Often called "the first lady of engineering," Gilbreth was cofounder with her husband of the Gilbreth consulting firm, which she headed for fifty years after her husband's death. She identified four skills as necessary to succeed in management:

- Knowledge of individual behavior
- The theory of groups
- The theory of communication
- A rational basis of decision making[10]

There is no indication that the situation has altered in modern corporations. If anything, recognition of the need for good communication has increased. Witness the growth of in-house workshops and developmental programs for executives. Many of these are directed at improving various communication skills. Perhaps this need for personnel, including technical professionals, to be communicatively competent is what the executive editor of *Industry Week* had in mind when he wrote:

Ironically, the management of technology is becoming more and more dependent on the human hand in both its delivery and its acceptance. . . . Unless management can deal with the human questions, as well as master the problems of technology, it will be fighting a losing battle to improve its products or productivity.[11]

The "human hand" in technology involves the effective sending and decoding of messages. People who have this ability deal with others in a manner that increases efficiency and productivity. Their ideas gain acceptance when there are differences of opinion. They get people to work together with a minimum of bad feelings. They have credibility, and their presence in the corporate world is vital. Further, it is the communicative skills such people have that make them seem so believable. Studying people with a Master of Business Administration degree (M.B.A), John Costello concluded that skill in oral persuasiveness ranked as the most important skill in business.[12]

Success

It is an overworked statement to say, with Costello, that communication skills are a necessary condition for success. Nearly every how-to book lays the fault of nearly every human problem at the feet of the "lack of communication." As with all half-truths, some of the idea makes sense, but it is a cruel trap to claim that competence in communication will propel the new professional to the top. Sometimes a person of rare talent, an Einstein for example, is a poor oral communicator, and it makes no difference. For those of us with more usual gifts, however, comparisons with our peers necessitate examining how well we are able to deal with coworkers, gain acceptance for our ideas, and articulate the problems of our research. In a recent interview at Essen, West Germany, Hans Juntgen of Bergbau-Forschung, a mining research firm, said, "We look for people who speak their ideas convincingly, for they reflect well their company also."[13]

The managers of modern corporations assume that those they hire have subject-matter competence. As a result, the qualities that distinguish one person from another extend beyond how much a person knows. We believe the cutting edge of success is communication skill. Sending those messages that speak of a *positive you* is the subject matter of the following chapters.

ACTIVITIES

1. Keep a diary for twenty-four hours. Note how much time you spend:
    a. In intrapersonal communication (thinking, dreaming, muttering, and so forth).
    b. In listening to another. Note what the *other* is doing while you listen (playing an instrument, speaking, singing, and so forth).
    c. In dyads. With whom? For what purpose? With what results?
    d. In groups of fewer than five persons, each sharing the re-

sponsibility for the communicating. Why did you all come together? What were the results?

e. In watching TV or engaging in some other form of mass media.
f. In reading and/or writing.
g. In groups of more than two when *you* have the prime responsibility for the communicating.

Note the proportions of time spent in each activity, how many activities are done simultaneously, and what your findings tell you about how you spend your time.

2. Project yourself into a professional position ten years from now. How do you think the proportions of time spent in each activity in question 1 would change?
3. Read an autobiography of a top executive, especially one who directs a company where significant numbers of technical people work. Assess:
   - What skills does that executive value in himself or herself?
   - Where and how were these skills learned?
   - What skills does that executive demand from corporate leaders and subordinates?
   - Where does human communication fit into the skills the executive expects from those in various company positions?

NOTES

1. Paul Tory Rankin, "The Measurement of the Ability to Understand Spoken Language" (Ph.D. diss., Univ. of Michigan, 1926), *Dissertation Abstracts* 12 (1952), 847–48.
2. Andrew D. Wolvin and Carolyn Gwynn Coakley, *Listening* (Dubuque, Iowa: Wm. C. Brown, 1982), 4–5.
3. Wendell Johnson, *Your Most Enchanted Listener* (San Francisco: International Society for General Semantics, 1956).
4. Gini Kopecky, "Betrayal: Between Friends," *Redbook*, Sept. 1983, 108; Cynthia Heimel, "How to Flirt," *Cosmopolitan*, Aug. 1983, 146; Robert Masello, "How to Seduce a Man," *Cosmopolitan*, July 1983, 90; Ruth Bell and Leni ZeigerWildflower, "Kids' Coded Messages," *Ms.*, vol. 12, no. 1 (1983), 42–43; Norma Peterson, "How to Stop Fighting and Start Loving Again," *McCall's*, vol. 110, no. 11 (1983), 8. Popular magazines such as these or the many advice columns in daily newspapers reflect the persistent need people have to establish a sense of control over their relationship with some other individual person.
5. Barbara Gastel, *Presenting Science to the Public* (Philadelphia, Penn.: ISI Press, 1983), 122.

6. Marianne Kyriakos, "The Hiring Game: Trimming the Odds," *Modern Office Technology* 29, September 1984, 62–68.

7. Darrell Sifford, "Duke Is Seeking Would-Be Doctors with People Skills," *Philadelphia Inquirer*, Monday, December 24, 1984, Sec. C, 8, col. 5–6.

8. "Education in Industry: Synopsis of the Joint ECAC-RWI Feedback Committee Report," *Journal of Engineering Education* 55, May 1965, 254–56.

9. Gerald M. Phillips, *Communicating in Organizations* (New York: Macmillan, 1982), 15.

10. Lillian M. Gilbreth, *The Psychology of Management: The Function of the Mind in Determining, Teaching and Installing Methods of Least Waste* (New York: Sturgis and Walton, 1914), quoted in Martha M. Trescott, "Women Engineers in History: Profiles in Holism and Persistence," in *Women in Scientific and Engineering Professions*, ed. Violet B. Haas and Carolyn C. Perrueci (Ann Arbor: The University of Michigan Press, 1984), 181–203.

11. Perry Pascarella, "Human Strategies," *Industry Week* 6, June 13, 1983, 7.

12. John Costello, "What You Need to Climb the Business Ladder," *Nation's Business* 64, June 1976, 6,8.

13. Interview conducted by the authors with Dr. Hans Juntgen, Bergbau-Forschung, Essen, West Germany, May 29, 1984.

CHAPTER TWO

COMMUNICATION MODELS

To develop an understanding of communication, you need a common vocabulary and grasp of the process. Fortunately for those trained in technology, model making provides a familiar means toward that understanding. Scientists have always relied on physical models of the concepts they explained. Some models are three-dimensional, and they help us to understand things that would be hard to visualize in any other way. Many physics students wrestle with concepts such as the structure of matter and find them elusive. In this case, the Bohr atomic model helps the student visualize the concept. When Jennifer sees a marble representing a nucleus and another single marble representing the electron as it spins around the nucleus, she forms a mental picture of the hydrogen atom. Yet, any student must realize that the Bohr model is not really like an atom. The structure is one expedient way of symbolizing a few complex concepts, but it also seriously distorts others. If these aids and shortcomings are recognized, models can be worthwhile learning devices.

Most models are not three-dimensional. Science texts are filled with print diagrams that serve the same purpose. The solar system model is often represented two-dimensionally. Think how difficult it is to understand the principle of electrical current flow until you look at a circuit diagram which presents that phenomenon as a rectangle with a few symbolic parts. See Figure 2-1. The theory becomes manageable as the picture becomes familiar.

Model making is an effective way to present abstract and hard-to-follow ideas, especially to those with scientific or technical backgrounds. We plan to capitalize on your orientation by building a diagram model of communication. We think you will find it a comfortable way of understanding some of the new vocabulary that describes the process.

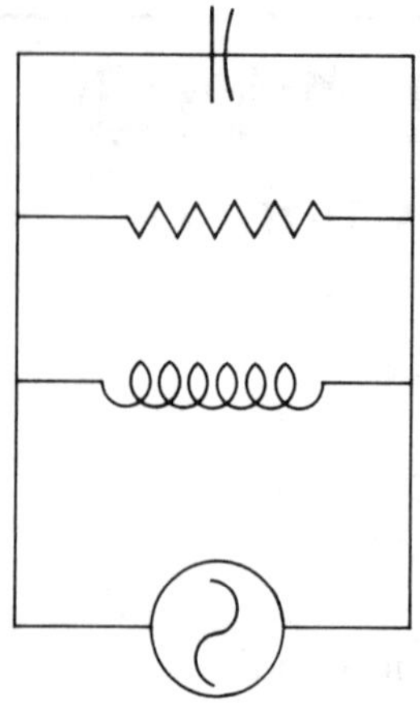

Figure 2-1 Diagram of simple circuit.

Communication models are not new. One of the earliest ones is the Shannon-Weaver model, presented nearly forty years ago.[1] Shannon and Weaver were not talking about human communication at all; but early theorists saw the application possibilities, and the model became widely used. See Figure 2-2. By today's standards, the model is inaccurate. Many others that have been generated reflect the growing understanding of communication.[2] We have chosen to present you with an original model and discuss it part by part. Our

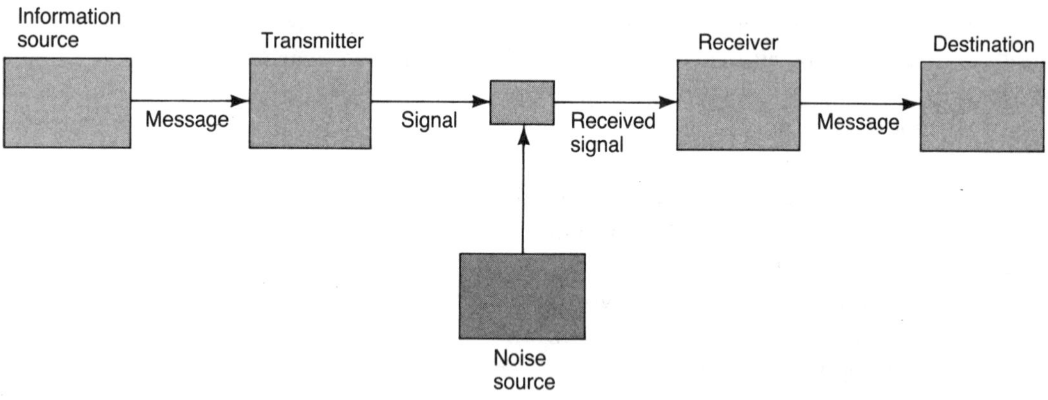

Figure 2-2 The Shannon-Weaver model. From C. E. Shannon and W. Weaver, *The Mathematical Theory of Communication* (Urbana, Ill.: Univ. of Illinois Press, 1949). Used with permission of the University of Illinois Press.

model does not deny the work of the earlier designs, but it does focus on those portions we find most applicable to you.

Perhaps we should consider some cautions necessary when using any model. This representation of the world is just that: a symbolic representation. It stresses certain things, but it also leaves out a great deal. It is meant to help you understand the communicative process. Use it to form visual pictures in your head, but never mistake any model for the thing itself.

THE CONSTANT VARIABLES IN COMMUNICATION

Perhaps *constant variables* seems a logical impossibility. After all, things are *either* constant or variable, right? Not exactly. We contend that certain elements are present in all communications, but these elements function in a unique way each time a communication happens. It is, for example, a constant to have some originator of the message, but who that originator might be is endlessly variable.

Think of communication as a transmission/translation process. Misunderstanding a broadly used term such as communication is not unusual. To many people, it means the exchange of oral words, a sense that seems to us too narrow by far. To others, all human behaviors are in some way communication, and that sense seems too broad. For expedience, assume that the process has a small number of parts. These parts, called constants, are always present, and so they can be identified and studied. Yet, though the parts are constant, the specific way they merge in every different setting changes. This transformation is the ever-shifting pattern of the process.

Suppose David wanted to send you the message, "I am sad." How might David transmit that message? He could

- say the words, "I am very sad."
- say other words, but in such a way that his voice conveys sadness.
- write the words.
- say nothing, but express sadness through the posture of the body and the manner of movement.
- compose his face into a mask of sad feeling.
- cry.

These examples come readily, for we have all used and seen them. Yet, there are many other ways to send the message of "I am sad," and they are equally familiar but somehow do not come to mind so quickly. David could also

- dance the message of sadness.
- compose a piece of music that, when heard, calls up the feelings of sadness.
- create a painting or a statue that, when viewed, is recognized as sadness.

All the preceding choices are available when David sends you this one message. That there is always a choice is a constant, but its selection varies almost endlessly. It is time to examine these constants and build the model of communication for our use.

The Sender

The one who starts the communication on its way is its *sender*. The sender is the originator, the beginning point, the initiator. Computer scientists call this source the encoder. Picture the sender as shown in Figure 2-3. It is not correct to restrict the concept of the sender to humans. Recall from Chapter 1 that barn swallows, cardinals, cats, and other living creatures send messages also. Humans are sometimes part of the communication, and humans will be our focus in this book, but keep in mind that other creatures are part of the communication process as well.

Does the sender have to be a *living* being? Most of us can accept the notion of dogs and elephants sending messages to others, but what if the sender is inanimate? Can a wire flapping against the side of the house or the wind rattling loose windows be considered a sender? Yes. These things do start messages on their way, so keep in mind just how these sounds (and sights) function. The confusion arises when someone *uses* an inanimate thing as the means through

Figure 2-3 The communication model begins with a sender.

which to send a message. Your little brother might be trying to frighten you by rattling that window, for example. In this case, the window is not the sender, but simply the means through which your brother reminds you of the horror movie you saw last night. For now, keep in mind that the sender is the point of origin.

Background

Senders do not materialize fully formed out of nothing. Each comes to a moment of a particular communication with a complete set of experiences. We call *everything* that has happened to the sender up to this moment the *background*. See Figure 2-4. As the model shows, the background serves as the foundation for the sender. Depending on the variables within the background, the sender's choices vary. If the sender is a person, factors such as where she grew up, where he went to school, what language was spoken at home, whether she learned to play baseball, whether he learned to cook all serve as agents to shape the sort of person who is at the communicative moment in question. If you have never studied art history, references to impressionism and cubism probably do not figure in your conversation. Those required college courses you take outside your major discipline place facts and understanding in your background so that you can choose from them in future communications.

Other creatures have backgrounds too. Your dog responds to commands or ignores them depending on how you introduced them in the dog's past. Animals, like people, have lifelong histories, and

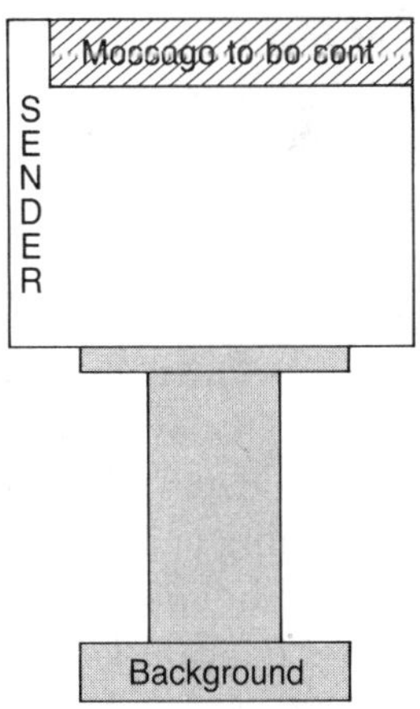

Figure 2-4 The background represents everything that has happened to the sender.

those histories constrain what is possible for the creature to choose at any given moment of communication.

In an analogous manner, windows and wires have a background too. If there is no solid weather stripping around the frame, or if a nail comes loose, the window rattles. If the wire has not been secured, it flaps against your house on a stormy night and makes you wish you had not watched that movie. Good communication requires that you consider the background of the given sender. Some of the surprising reactions we get to things we say or do might be the result of what was present in the background. Notice that the *message to be sent* derives from this background. It is unlikely that a sender, animate or inanimate, will inaugurate a message unless something in the sender's past makes the choice of that message possible. We do not mean to imply that the sender cannot create novel messages. From infancy, humans generate ideas and verbalizations new to them. What is impossible is that these messages contain input data totally absent from a person's background.

Stimuli: Random and Ordered

We have used the term *message* loosely so far. Technically, a message can never be sent. Messages are formulated in the sender, but what is sent is some set of stimuli—some signal, as the Shannon-Weaver model calls it. A *stimulus* is a bit of activating energy. Disturbances in the air waves that transmit as sound or patterns of light waves that are sent on as sightings are stimuli. See Figure 2-5.

Understanding of communication was retarded for years when theoreticians believed that words or thoughts (or messages) themselves were sent. This view made it possible to visualize the sending as a sort of full package of some ready-made transmission. The package was formulated and the destination received it *as is*. We now believe that view was erroneous. What is sent forward is an energy source with the potential to be made into messages.

Stimuli as they occur in communication are of two sorts: random and ordered. *Random stimuli* occur without regularity and structural agreement. When someone knocks over the trash can in the hall, the stimulus of the sound waves allows you to perceive that the can has fallen again. Those wires flapping against the house and the rattle of the windows are examples of random stimuli. They are clearly communications, but they have no rule-governed order. Nobody has agreed that *each time* a wire or window makes noises we are expected to take it as having some specific meaning.

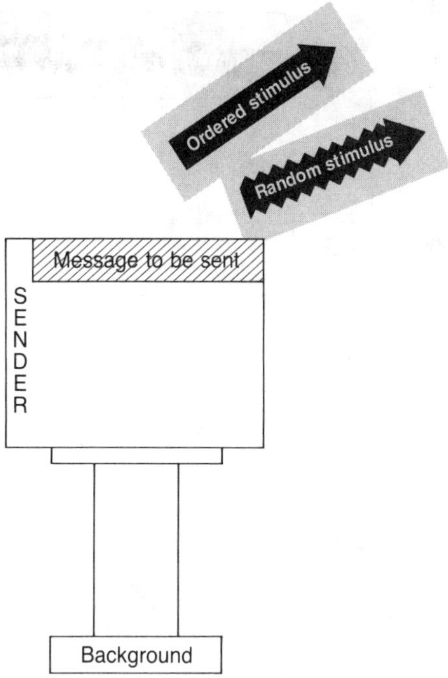

Figure 2-5 Stimuli can be ordered or random.

Some sights and sounds *do occur* according to the agreement of the user's community and a rule-governed structure. That is, people in the community *decide* that something is going to count as meaning something else. These are *ordered stimuli,* and they are usually called *language.* When sounds are made according to predictable rules, they form themselves into symbol systems of words, sentences, and paragraphs. Even single-sound patterns, such as a fire siren code, become rule-governed bits of language. The key to differentiating language from, say, nonsense noises, is whether the symbols have agreed-upon rules associated with them. If they do, the symbols function as language.

Not all language is verbal. Many gestures, facial expressions, spatial arrangements, colors, objects worn or used, and so on, function as language systems agreed upon by the community. Their order is predictable, and some theorists call such stimuli nonverbal languages. We shall have more to say about languages in Chapter 3.

Channel

The *channel* in communication is the choice available to the sender through which the stimuli will be sent. In Figure 2-6 we have repre-

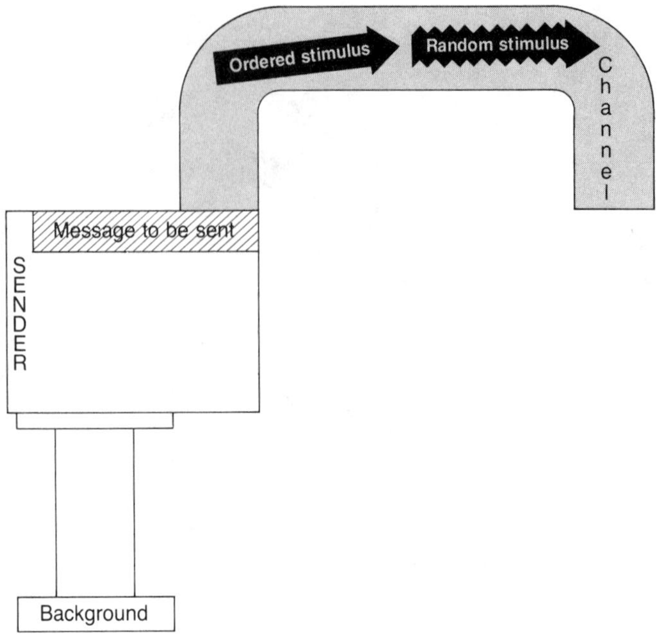

Figure 2-6 The channel is the choice selected for sending the stimuli.

sented the channel as a pipeline through which these stimuli pass. We were describing channels when we spoke of the ways David had to send the message, "I am sad." Each of us selects thousands of channels each day as we decide to telephone rather than write a letter or speak to our friend directly rather than leave her a copy of the song written just for her.

Perhaps you can see already that each channel has its own special constraints. My friend can hear my song in the dark, but I cannot show her the painting unless there is light. The blind person could not read my letter, but he could feel the sculpture I crafted. The loud music at a mixer does not prevent students from communicating their interest or lack of it to another. Not much talking goes on, but the stimuli are sent along channels where talk is not important; and the mixers seem to get along just fine.

The Receiver

The *receiver* accidentally or intentionally is the recipient of the stimulus. The receiver forms the symmetrical parallel to the sender. If sender acts as the originator, the beginning point, the initiator, or the

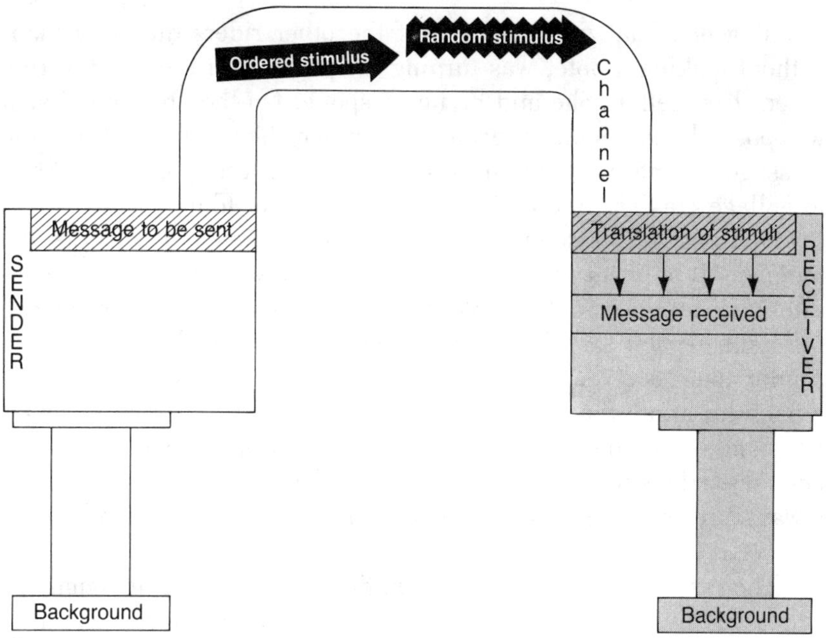

Figure 2-7 The receiver is the one who decodes the stimuli.

encoder, then the receiver could be called the terminal, the end point, the concluder, or the decoder. See Figure 2-7. Receivers and senders are similar in many ways: they can be human, nonhuman, or inanimate; they come to the moment of communication with a background that, largely, structures how they will be able to receive the message.

Consider how many receivers we have each day. Some are the ones we have chosen to receive our stimuli, but many are accidental participants in some communication. Each of us overhears conversations, catches sight of someone doing something, or finds letters meant for another. Museums are filled with objects seen by thousands but that were painted or crafted by an artist for only one other person.

One of us was the accidental receiver of a message that she has never forgotten. She used to return from university classes by bus at the time that a large auto manufacturing plant in Detroit was changing shifts. She noticed one of the workers, a middle-aged man who seemed no different from any of the others, except that he took a book out of his lunchbox and read all the way home. She was intrigued because the book was an anthology of Aristotle's works. Instead of sleeping or socializing, as most of the other riders did, this man, without making a note, was turning the pages as if the book were a recent best-seller. She and he never spoke. On the crowded bus, it was possible that he never even realized that she was present, but the message she interpreted him as sending was: Learning is not confined to college classes. She has never forgotten it or him.

One possible misconception might arise when you consider the receiver. It appears as if the sender and receiver must be two separate beings; that is, two *different* individuals serve in these two different capacities. That is not accurate. For people at least, the sender and receiver might be the same person. All of us talk to ourselves, argue the pros and cons of this or that, try out conversations and revise them, or, in short, act as our own receiver. That most enchanted listener we spoke about in Chapter 1, that one who is always present when we say or do anything, is the self acting as receiver.

The essential difference between the receiver's role in communication and the sender's is that the receiver has to *translate* the stimuli that the sender dispatched. Remember all that the receiver has to work with is the raw material of sensory stimuli. The receiver must operate on them, order them, and translate them into a message that comes close to the message that was forwarded. Sometimes this translation function is called listening or perceiving. Each of us has stories of how messages we sent or received were interpreted badly. Some of these might be humorous, but others can be frustrating indeed. As we become aware of how the sender's message and the receiver's interpretation of it differ, we might begin to recognize some of the potential pitfalls of misunderstanding.

Feedback

Feedback functions as a monitoring device, much as a thermostat does in a home heating system. In communication, feedback is the reaction to the stimuli that the receiver sends back via other stimuli to the sender. Feedback lets the sender know how the message was decoded. See Figure 2-8. As with the original stimuli, some of the feedback is *ordered* and some is *random*. A person speaking before a large audience, for example, watches for feedback reactions. If you were the speaker and told a joke, you would wait to see if anyone laughed. The laughter is your feedback, and you would probably be encouraged to keep that joke for use again. The same is true of applause and it, like laughter, is an orderly and rule-bound feed-

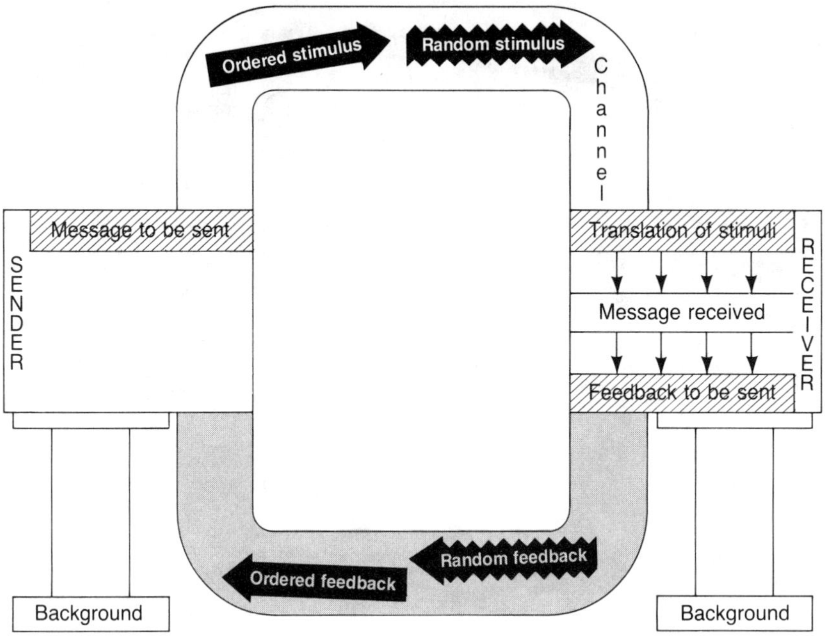

Figure 2-8 Feedback is the receiver's reaction to the message.

back. However, you might notice that someone in the audience yawns. The listener may be extremely interested in what you are saying, but a momentary and random need for oxygen promotes the yawn. You, seeing the yawn, would probably translate its feedback as boredom.

The feedback portion of the model was missing from the work of the early model makers such as Shannon and Weaver. At that time, communication was considered a one-way process, and the direction was from sender to receiver, where the communication was assumed to end. Current theory recognizes that the process comes back on itself: it moves from sender to receiver, where some reaction takes place that is, in turn, transmitted to the sender. The communication does not *end* at this point. It feeds forward to communications and behaviors in the future. In this circularity and feed-forward concept is the assumption that each part of the process affects every other part. Technically trained persons should recognize this analysis as a systems approach, where every part of any system is assumed to have a cause and effect relationship with every other part.

The sender, paralleling the activity of the receiver, must both *receive the feedback* and *translate* it, as shown in Figure 2-9. In

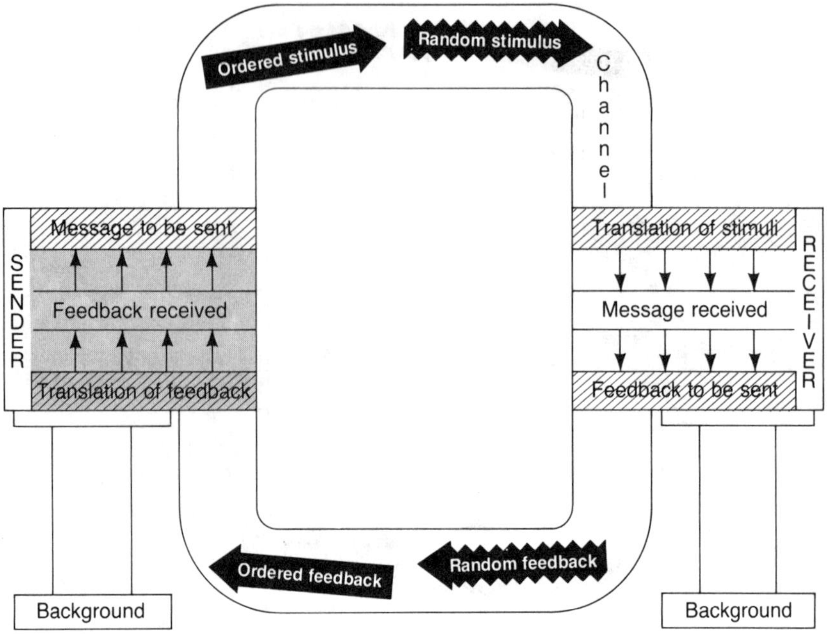

Figure 2-9 The sender receives and translates the feedback.

essence, the sender learns how the original message was translated by the receiver through the stimuli that the receiver feeds back to the sender. As with machine monitoring, this feedback should then be used to structure the sort of message the sender transmits next.

Consideration of feedback is an essential portion of understanding communication. It acts to tie the sender and receiver together in any communicative act. Actions that block feedback or refuse to take note of it should be avoided. Every stare, yawn, and receiver movement should alert the sender that the next set of stimuli sent might have to be altered. One of the corollaries of the following chapters is to learn how to interpret the feedback of receivers so that the communication might proceed at its optimal level.

Perhaps it is obvious that some feedback *encourages* the next message, and some *discourages* its continuation. Feedback that encourages is called *positive*, and feedback that discourages is called *negative*. When the joke fails to bring laughter, you are discouraged from keeping it in your "bag of jokes to tell." When the joke is good, but your hearer reacts to tell you that this is the wrong time to tell *any* joke, you are likely to make the rest of your remarks less humorous. In both instances your subsequent message choices were altered through the monitoring of negative feedback. When a receiver laughs when you anticipated it, appears to be listening intently, or makes agreement sounds and movements, you are positively encouraged to proceed as you have been doing. If you fail to take note of what the receiver is telling you, you lose the monitoring input.

Entropic Noise

From science comes a convenient concept to explain why communication goes astray. Communication theorists have called the concept *noise*, while scientists call it *entropy*. Both seem to be talking about the same thing: the tendency of events to fall into disorder. Among the many interpretations that exist, two statements about *entropy*, the first taken from a high school chemistry book and the second taken from a college text, are useful explanations:

What we observe here [when an ice cube melts] is a tendency for the ice to move into the more disordered liquid state. This suggests a second basic rule for natural processes: the tendency for processes to occur which lead to a more disordered state. *This tendency in nature is toward greater randomness in a system.* That property which describes the disorder of a system is called entropy.[3]

The direction in which natural processes take place (toward higher entropy) is determined by the laws of probability (toward a more probable state). The equilibrium state is the state of maximum entropy thermodynamically and the most probable state statistically.[4]

The process of communication also functions at equilibrium in a "state of maximum entropy." This inherent inability to move stimuli from sender to receiver without losses has traditionally been called *noise*. A. Bruce Carlson, in his book *Communication Systems*, says that, "Broadly speaking, any unintended signal perturbation may be classified as 'noise,' and it is sometimes difficult to distinguish the various offenders in a corrupted signal."[5] He cites these offenders as contamination, distortion, interference, and noise. Since the concept of entropy is probably familiar to you, it will be used throughout this book to speak of messages tending toward disorder. See Figure 2-10. The model suggests that entropy pervades each part of the communicative system. Consider the potential for losses whenever any message is (1) constructed by the sender and (2) forwarded in the form of stimuli through some channel to a receiver who (3) translates the stimuli into a message and then (4) reacts to it by selecting stimuli to feed back to the sender. We claim that it is impossible to transmit any stimulus without running into disorder. What you begin with in communication is never exactly what ends up in the decoding. Input does not equal output; message to be sent is not message as received; whatever goes in comes out altered.

Sometimes the entropy is masked well or is at such a minimal level that we might forget it is present. The masking might be due to poor feedback. When you pretend to listen intently to your Aunt Mildred's recitation of what's wrong with the world, she is encouraged to continue her recitation. You translate her message as unacceptable boredom, but you tell her by your feedback that her message is arresting and of interest. The loss of the message she *thinks* she is transmitting is considerable, and she probably never will know it. At other times the entropy might be held to a minimum. Your boss says, "Open that door," and you do. The communication seems completely accurate, because the very door that the boss wanted opened was, in fact, opened. Yet consider the losses: Do you know why the boss formulated the message about the door in the first place? Do you know how far the door is to be opened? Do you know exactly how the

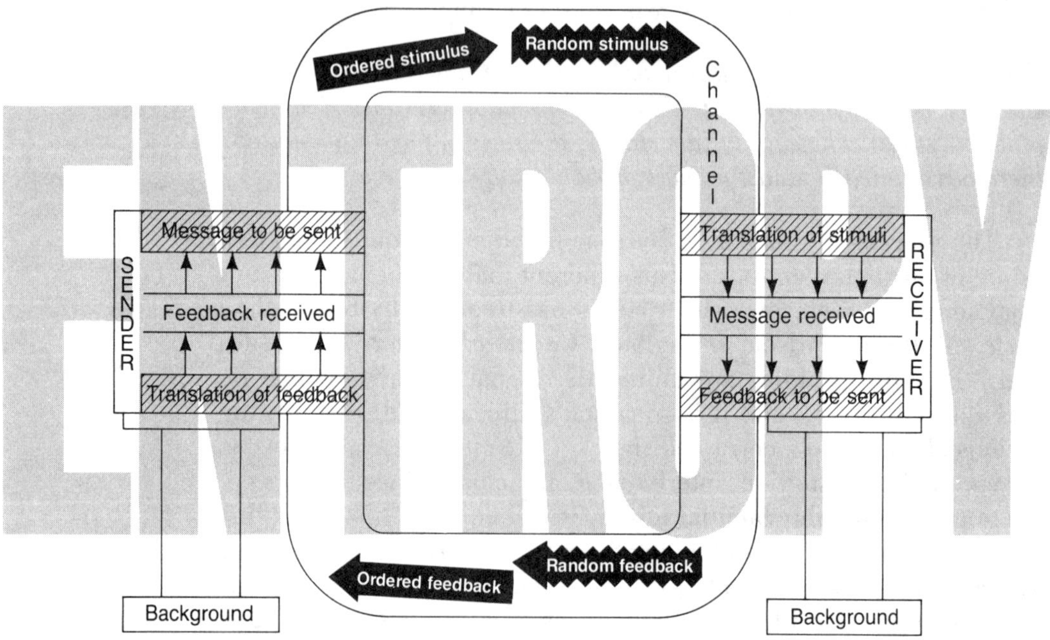

Figure 2-10 Entropy is the tendency of messages to fall into disorder.

boss feels about the opening of the door? Do you know how the boss feels about everything at the time of the request? Of course you cannot, yet all of this is part of the background out of which your boss, acting as sender, constructed the message. In reality, the best we can promise in communication is a communication in which entropy is held to its minimum so that a *fairly* accurate translation of the original is possible.

But how about the case of the same person acting as both sender and receiver? Surely that is an instance of complete transmission with no losses, right? Think about it. When you argue with yourself, how does the process go? You at time 1 say, "I'm going to a movie tonight." You, now at time 2, answer, "That's an awful idea. I have a calculus test tomorrow." You at time 3, respond, "I can do both. The movie's not long." Notice that *you* are moving through time as you say these things. You are in process, changing physically and mentally. You take on ideas about tests and movies, discard them, and change them. The you at the beginning of the dialogue is a very different person from you at another time. The changes might be minimal, but they are changes nonetheless, and the loss is inherent.

Entropy is an extremely important consideration. When a communicator accepts its presence, it is not nearly so disturbing an experience when some message is received in a garbled manner. When you know it is there, despite your best efforts to keep it in check, you have an effective start on keeping the system operative.

Setting

The last constant in communication, completing the model, is *setting:* the environment where the communication takes place. See Figure 2-11. Anything that is said, written, danced, painted, and so on, takes place somewhere. This setting is not a trivial matter. Some messages would be highly improbable in one setting, although they might be highly acceptable in another. One does not tell stand-up-comic jokes at a funeral eulogy, and neither is it likely for anyone to preach a funeral eulogy during a stand-up-comic routine. The kind of communication you select as appropriate when you are with your friends in the gym is not necessarily the kind you choose when you are in your grandparents' living room.

In Chapter 3, we will discuss various nonverbal and verbal communication patterns. Objects, colors, smells, and space arrangements are part of these nonverbals; words form the verbals. All are also a part of setting.

One pleasure of human rituals like wedding ceremonies or graduations is the contrast in the sort of messages spoken at such times and the sort spoken at other times. Obviously, where you are at a given moment both constrains the sort of messages sent and is constrained by those messages. We, as a society, expect certain things to happen when we attend religious services; conversely, the nature of the religious service is structured by the sort of things that usually happen there. As with all the communication constants we have discussed, the setting helps explain why and how the various pieces of the communication took place. With setting, the communication model constants are completed.

The Model of Communication

With all the constants in place, we must realize that no communication exists separate from all other communications that preceded it. Neither is it separated from those that come after it. A person, a nonhuman animal, or a machine that communicates draws from everything that happened to it before and feeds forward to everything

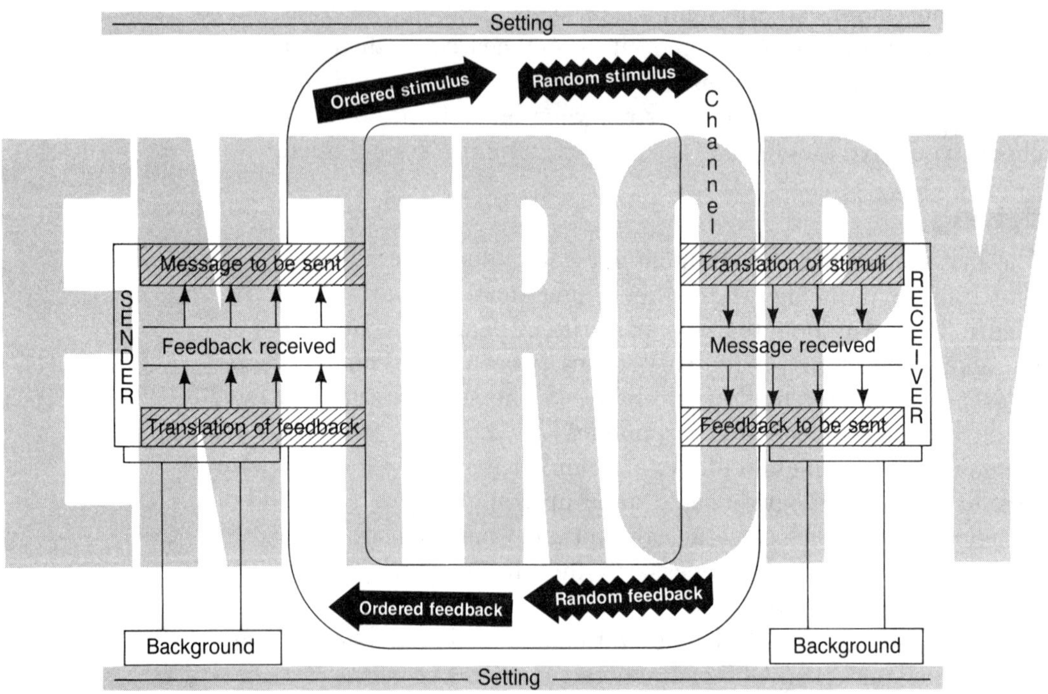

Figure 2-11 With the setting, the communication is complete.

that will happen in the future. We represent this link of past and future in the form of a helix. See Figure 2-12.

The figure reminds you, through the pattern of a helix, that very real links exist between all communications you have had. These links form the spiral wherein your past (background) leads to the choices available to you in your future.

THE TRANSIENT VARIABLES IN COMMUNICATION

We have just finished arguing for recognition of a set of constants in communication. Looking for and dealing with such factors as sender, receiver, and so forth, help us understand the process. At the same time, we need to understand that each constant is transformed limitlessly. Senders are constant, but who they are changes without end. The same sender at two different instants is a vastly different person. There is always a channel in communication, but what it is and how many there are change with each communication. Entropy is inescapable, yet where it is strongest and where it has least effect alter from second to second and instance to instance. To understand communication, you must take each of the constants and subject it to the analysis of how it varies *this* time.

For our purposes, we examine three different sources for these transforms of the constant variables. When we have looked at them and matched them to the previous section, then we are ready to look again at the completed model of communication and consider a verbal definition.

Experiential Effect

We have used the background of the sender and receiver as the foundation upon which these two constants are built. Perhaps you already see how experiences shape these segments. If we consider experience as behavior that took place in the past of some creature, it is simple to recognize how the nature of the experience structures the remainder of the creature's communications. It is unlikely that the ordered stimulus of the Russian language would be chosen by a Spanish sender of an everyday conversational message. The experience of learning Russian is probably missing from the Spaniard's background. It is likewise improbable that a modern American military unit would select to dance a given challenge message to its opponents. Dance, as part of warrior training, is not in our back-

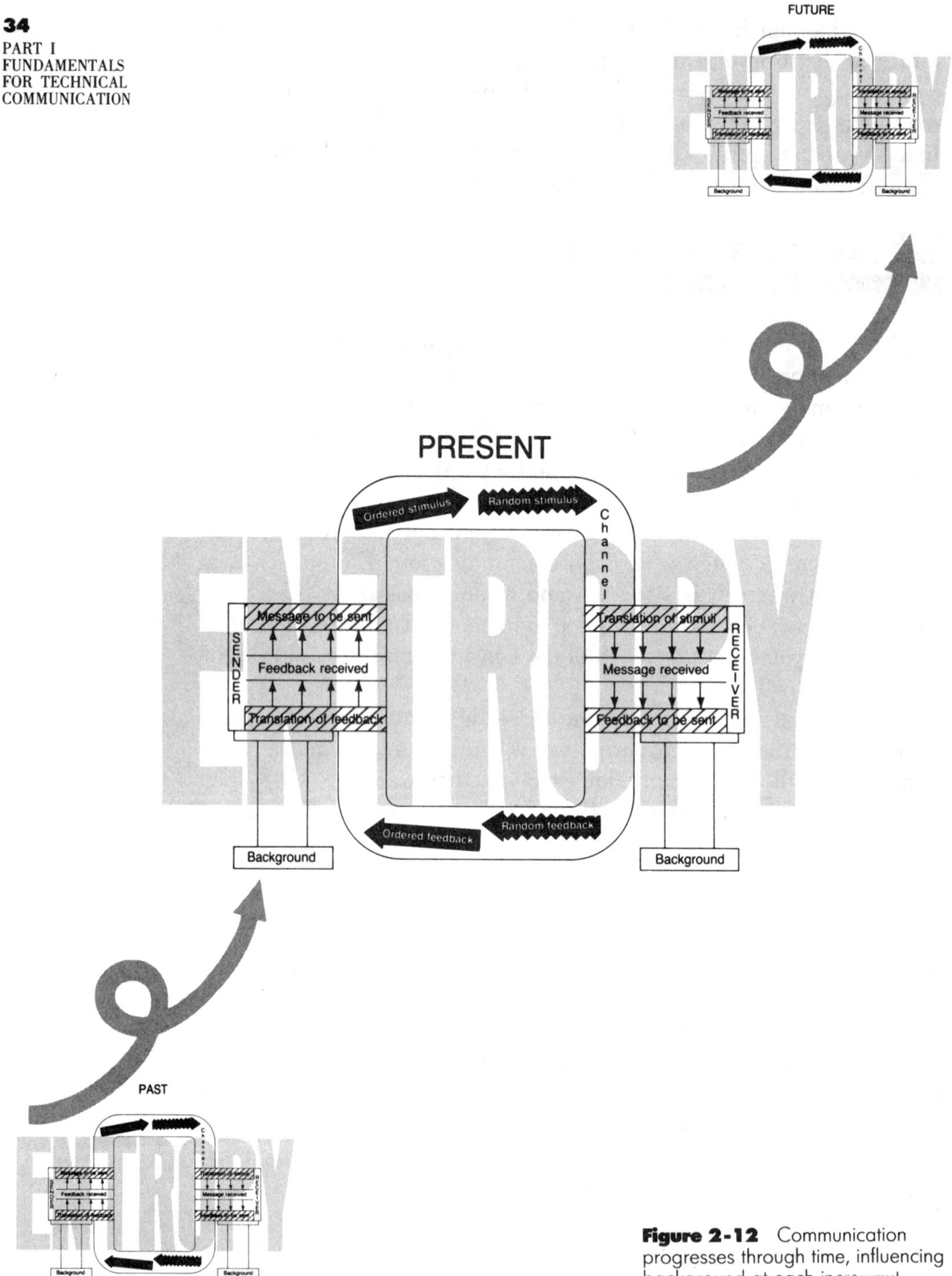

Figure 2-12 Communication progresses through time, influencing background at each increment.

ground. Of course, if the Americans were sixteenth century Indians, the probability would increase that the dance channel might be selected.

One reason most of us are uncomfortable in unfamiliar situations is that we have no previous experience to guide our expectations. If you had never eaten in a formal restaurant where multiple pieces of silverware are used and many unfamiliar place settings of dishes are served, you would probably be very apprehensive. Many of us do not know which of the numerous forks is supposed to be used for the oysters and which for the pâté. Therein lies the fear. Not only would it be pure chance for us to choose the right one, but many of us would also avoid such a restaurant setting entirely.

Experiences give us a way of organizing our thoughts about something that has not yet happened. At any particular moment, the constants in a communication will vary according to the experiences that preceded the communication. Knowing what those experiences were gives us some predictable expectation of what the constants will be like this time.

Perceptual Bias

Each event in this physical world can only occupy a given space at a given time. Anything that is perceived from a certain space is not accessible to any other being at the same time. In short, at this moment, only those creatures who are in some one place are capable of taking in some perceptual object from that space. When we are at a moment of communication (and it, of course, takes place in some setting), only we are able to perceive something from that exact spot. Where we are, in part, determines what we are able to perceive *and our exact perception is closed off to anyone else.*

What we are talking about regarding *perceptual bias* is another way of thinking of the Six Men of Indostan story. If you remember, the story tells of six wise blind men who bump into an elephant. One touches the elephant's side and claims the beast is like a wall. The second feels the tusk and assumes the animal is a spear. The third grasps the trunk and concludes it is a snake. The fourth feels the knee and decides it is a tree. The fifth chances upon the ear and decides it is a fan. The sixth touches the swinging tail and declares the elephant to be a rope. The point of the story is that they were all right, for they described, in fact, what they perceived; but, at the same time, they were all wrong, for the elephant is none of these. See Figures 2-13 and 2-14.

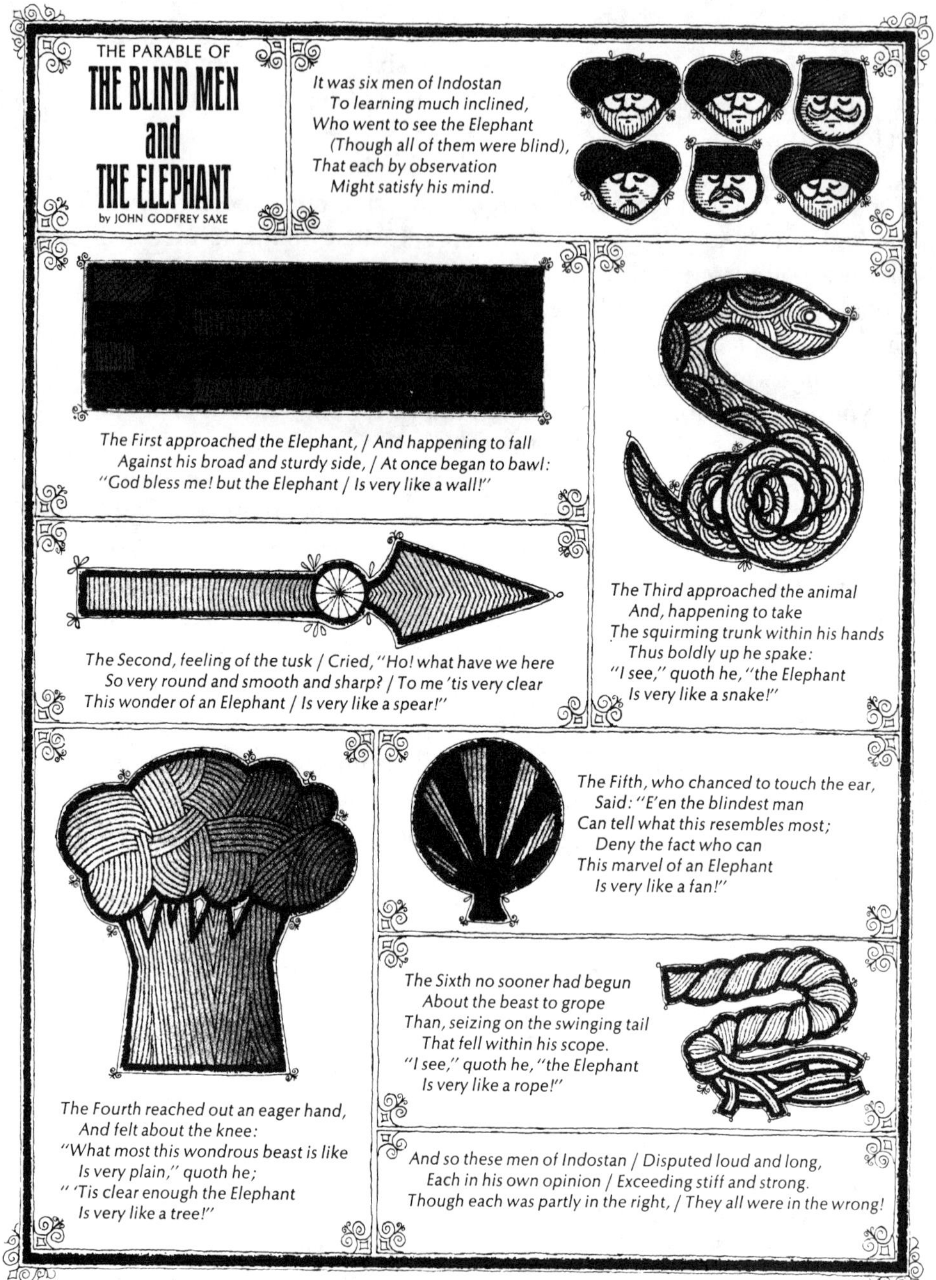

Figure 2-13 The story of Six Men of Indostan. From Don Fabun, *Communication* (San Francisco: International Society for General Semantics, 1977). Copyright © 1968 by Kaiser Aluminum & Chemical Corporation. Used with permission of Macmillan Publishing Company.

Figure 2-14 The composite elephant, as seen by the Six Men of Indostan. From Don Fabun, *Communication* (San Francisco: International Society for General Semantics, 1977). Copyright © 1968 by Kaiser Aluminum & Chemical Corporation. Used with permission of Macmillan Publishing Company.

If you remember that someone, no matter who he or she is, occupies a unique place in time and space, then you can see that a communication is not exactly reproducible. A similar occurrence might happen again, even to the point of the same stimuli sent by the same sender and translated by the same receiver as before, but it is a *similar* rather than an identical happening. The time is different, and the constants have altered. The perception of what might be happening is yours and yours alone. At this moment, nobody but you occupies your physical space.

The point about perceptual bias is important as we come to realize that reception is different for each receiver, no matter how many there are. When you sit in a large classroom with two hundred other students and listen to introductory lectures on biology, the teacher forms a unique communication with each of the students present. Look around such a hall. There is the "same" sender, using the "same" channel to send the "same" stimuli. Yet look at the differences the receivers make of those stimuli. The student next to you might be asleep, the one in the next row might be studying for tomorrow's test, while you might be listening with fascination. Only you live in your skin, and so only you know the translation process you are going through. Others might be perceiving something quite different from what you are, and the variances in the sort of message you each claim you heard could be significant. Surely that comes as no surprise. Perhaps you have had the experience of "comparing notes" after such a lecture only to find that what you thought was going on is not at all what others did. Compare "pop quiz" results after such a lecture. The grades indicate that translations of the message varied from some fairly exact decodings to some that were entirely erroneous.

Any group is made up of individual receivers, each re-creating the sender's message through a unique perceptual system. It is pointless to consider communication in terms of *audiences* or *classes* unless we remind ourselves constantly that what we really mean is that an audience or class is composed of separate and utterly varied individual perceivers.

Professional Expectations

The final element we would like to discuss in presenting the communication constants as variant is the effect of expectations your profession has. The kind of constant likely to appear in those communications involving the technical professional are part of the culture that makes up scientific communities. Any corporation is a society. It has a set of rules and rewards that may be quite different from those of any other corporation. Yet, among technological corporations (and many modern multinational companies *are* technological) certain industry-wide expectations are apparent. These expectations affect the way each of the constants in communication is likely to appear.

Technological professionals, as senders, are supposed to be competent in subject-matter background. With the need to specialize comes the expectation in modern corporations that a given person will have in-depth knowledge of some body of information. A recent graduate in heat transfer must know the factors affecting the flow of heat but must also be able to construct a computer-model simulation of heat flow. Those who send messages and those who receive them bring to their communications firm expectations that the others will have competent knowledge of the subject.

Assumptions are also made about how an "educated person" should sound and write. Faulty grammar in reports and sloppy articulation in spoken speech cast doubt on the professional's capabilities. Communities expect their college-trained people to be different from those who have never had advanced education.

Likewise, the stimuli chosen to be sent are presumed to be language systems containing the vocabulary of scientific terms. Nontechnologists often complain that, "Nobody can understand what scientists are talking about." That criticism is a reminder of an unfortunate oversight by a professional of any discipline. Outside the discipline the language constant varies significantly from the way it functions inside the discipline. Professionals are expected to keep the differences straight and know when to use which language.

The channels of communication in modern corporations tend to be a blend of the written, the spoken, and the visual. Each channel is bound by individual corporate expectations. Presentations, for example, are usually of a certain kind. In some instances, an oral review without transparencies is unthinkable. At a recent International Energy Conversion Engineering Conference, not a single paper was delivered without the use of slides or transparencies. The very nature of these corporate-wide expectations requires special chapters within this book to prepare you for corporate and professional presumptions.

When these intercorporate expectations are added to the special requirements every company develops as its unique "way of doing things," you begin to get a sense of how the constants in communication "undergo transform." It is never enough to simply identify, say, the setting, unless you also subject that setting to the analysis that reveals its special variance in the communication under discussion.

THE COMMUNICATION PROCESS MODEL

The communication process, as we have dissected it, consists of eight constants. Each of these constants, however, "undergoes transform" depending on the particular time when and the place where the sender transmits the message. Underlying all of the parts is entropy, and its randomness prohibits the totally accurate sending and receiving of any message.

A verbal definition of the model would be: *Communication is the sending of stimuli through some channel by a sender so that a receiver translates the stimuli into a message and then feeds back additional stimuli as a message monitor to the sender. In communication, as in the physical world, entropic noise is an inherent part of the process.* Communication always involves a finite number of constant variables that "undergo transform" each time they are used. To the degree that entropy interferes with the transmission, the communication is inaccurate. The aim of a good communicator is to keep the entropy at a minimum, so that messages can be translated with as little loss as possible.

Our model is presented as one way of considering how all communication works. It should be possible for you to take any sender-receiver exchange and label its parts. Try it for this case: You stand at the curb ready to cross the street, when you hear a loud horn blast. You jump back just in time to avoid being hit by a fast-moving car.

- *Sender:* Driver of the car.
- *Sender's Background:* Everything that the driver has experienced, including how to blow the horn.
- *Message to Be Sent:* Look out!
- *Channel:* Sound of the horn.
- *Stimulus—Random:* Non-rule-governed disturbances in the air waves.
- *Receiver:* You, the intended receiver.
- *Receiver's Background:* Everything that you have experienced, including the sound of car horns.
- *Feedback to Be Sent:* I hear you and am getting out of the way.

- *Channel:* Body movements.
- *Feedback—Random:* Disturbances in the light waves.
- *Feedback Received:* That person heard and got out of the way in time.
- *Entropy:* Minimal, since the translation of the message was quite accurate.
- *Setting:* A street you are about to cross while a car approaches rapidly.

Try another one this time: You enter the test chamber where a technician is completing work on a thirty-six-hour engine test. You are worried about the data output for the first twelve-hour period and say, "Let me see the data sheets for the first part of the test again." Technician hands you the output record for the first hour. "Not the first hour, Lou. I want the twelve-hour summary from the whole first part."

- *Sender:* You
- *Sender's Background:* Everything that has happened to you, including your worry about the test.
- *Message to Be Sent:* Lou, find and bring to me the twelve-hour summary of the first twelve hours of the test.
- *Channel:* Sound waves through which the words are sent and light waves through which Lou sees you.
- *Stimulus—Ordered:* Certain sounds arranged according to English agreed-upon rules.
- *Receiver:* Lou, the technician.
- *Receiver's Background:* Everything that has happened to Lou, including his knowledge of where the test results are.
- *Message Received:* You want me to find and bring the test results of the first hour.
- *Feedback to Be Sent:* I have found the test and am handing it over.
- *Channel:* Body movements.
- *Feedback—Random:* Disturbances in the light waves.
- *Feedback Received:* Lou did not know which test results I wanted.
- *Entropy:* Widespread. Problem with the sender, the choice words, possible fatigue of the receiver, lack of checking with the receiver.
- *Setting:* A test chamber at the end of a thirty-six-hour engine test.

There were multiple problems with the second communication, and the entropy was so pervasive that the original message sent forward by the sender was lost almost completely. The car and street-crossing instance functioned quite well, fortunately. If it had not, a serious accident might have happened. If you keep in mind that the same constants were operative in both examples, you will grow used to searching for them. If you also keep in mind that each constant is utterly unique for the particular communication happening at that moment and will vary in every other one, you have a start on understanding the process.

WRAP-UP

Our purpose in taking you through a detailed analysis of model building and how we assert that it works is a practical one. The model and its vocabulary give you the means of understanding the rest of the book. More importantly, an understanding of the process of communication gives you an analytic tool with which to examine what might be happening at the different communication points of your life.

Your career in the corporate world as a technical professional or as a person who works with such professionals will center on analysis. One of us heard a physicist explain to his class the difference between a technician and a scientist. "A technician has to be instructed what to do," he said. "A scientist figures out how to do it." Use the analytic training that is so much a part of science to figure out all those communication constants as they "transform" from one encounter to another. Such monitoring will help you to explain yourself and your particular entropies in all your interactions.

The model asserts an *interactive* process of systems analysis. No single variable is "at fault" when breakdowns occur. The entropy is always there. Both the sender and receiver must check each other to keep that entropy in control. If your technician seems tired as you give directions for how the test is to be run, you bear the responsibility for recognizing that tiredness will increase entropy and that your message might go astray. The technician also should let you know that fatigue is inhibiting the effective translation of stimuli at the moment. As people begin to understand what happens when message systems are crafted, they improve their interaction with each other; and corporations, as well as individuals, benefit.

ACTIVITIES

1. Recall any verbal conversation you have had in the last week with one other person. Identify the "communication constants" of sender, receiver, background, random and ordered stimuli, channel, message to be sent and message received, feedback, setting and entropy during this communication. Note how these constants shifted as the conversation developed.
2. Examine any issue of controversy on your campus, perhaps one reported in your campus newspaper. On the assumption that two (or more) people have different interpretations about what did or should happen, identify possible sources of entropy that might have caused these different interpretations.
3. Present a 3–4 minute presentation to the class in which you advocate some position you think the listeners should accept on some issue. When you finish, administer the following questionnaire to each member of your audience:

   - What were the two main points (or one or three) of my talk?
   - How do you feel about the issue I talked about?
   - Did my speech change your basic position about what you felt?

   Look at the differences among the answers you get. What kinds of entropy account for such differences?
4. Observe some student group meeting on campus. A student government or co-curricular activity meeting would be ideal. Note:

   - Does everyone agree with everything the chairperson says?
   - What accounts for the entropy?
   - Does the meeting setting cause any changes in the sort of communicating that takes place?

5. Tell four persons who know you well and are gathered together in the same place some piece of "confidential information" about yourself. Twenty-four hours later, talk to each of these four persons separately. Ask them what they thought you were saying about yourself and why you told them. How do you account for the differences in their interpretations?

NOTES

1. C. E. Shannon and W. Weaver, *The Mathematical Theory of Communication* (Urbana, Ill.: Univ. of Illinois Press, 1949), 98.

2. See, for example, the contrast in Raymond Ross, *Speech Communication Theory and Practice* (Englewood Cliffs, N.J.: Prentice-Hall, 1965), 1–19 and the sixth edition of the book published in 1982.

3. H. Clark Metcalfe, John E. Williams, and Joseph F. Castka, *Modern Chemistry* (New York: Holt, Rinehart & Winston, 1970), 43–44.

4. Robert Resnick and David Halliday, *Physics:* Part I (New York: John Wiley and Sons, 1977), 558.

5. A. Bruce Carlson, *Communication Systems*, 2nd ed. (New York: McGraw-Hill, 1975), 4–5.

CHAPTER THREE

THE SYMBOL SYSTEMS OF COMMUNICATION

No one part of the communication model is more or less important than any other part. As we insisted in Chapter 2, the model pictures a process, and its parts form a system. The body of knowledge known as system theory teaches us that consideration of any event as a *system* requires examination of each piece of that event as it affects and is affected by every other piece. So it is with communication. The central portion of the model represents stimuli in transit to the receiver and then other stimuli fed back to the sender. In Chapter 3, we examine these stimuli. We plan to look at both random and ordered stimuli; discuss how stimuli become symbols, which in turn become language; and look at both verbal and nonverbal languages. These topics are vital because they, like all parts of the model, are related to all the other parts.

RANDOM AND ORDERED STIMULI: SIGNS AND SYMBOLS

We have chosen to call the raw materials that pass between sender and receiver random and ordered stimuli. This differentiation is crucial, since many things communicate without involving language. When you hear the wind blowing hard and the sky darkening, you assume that it is going to rain. Neither the sound of the wind nor the sight of the sky is a language function, yet each works as a stimulus and communicates something to you. A stimulus is a familiar variable to the technologically trained. Another way to talk about the same concept would be to call it by its other name: sign and symbol distinction. A *sign* can be thought of as a natural and relatively random happening that holds communicative potential. A *symbol*, on the other hand, is created by humans to indicate something else.[1] The color white is a sign, but it is a symbol when we say it means *purity*. A diamond ring is a sign that becomes a symbol when it comes to mean an *engagement*. Your mother's tears at your graduation are a sign. You probably consider them a symbol of her *love*.

In each case, the essential difference between a sign and a symbol is that a sign happens randomly. The color could have been green or yellow, the ring might have been a bracelet, and the tears were just as likely to be a smile. Signs tend to happen spontaneously, even subconsciously. Nail biters are not likely to be aware of their activity at a given moment. We who observe that behavior, however, probably assign to it the symbolic meaning of nervousness. When we talked in Chapter 2 about the driver of the approaching car who uses the sound of a horn to tell you that you must get out of the way, notice that the sound of the horn was a sign, but you decoded it as a symbol meaning *warning*.

We are surrounded by signs. They happen haphazardly, without order or structure. For our purposes, recognize that these signs act as sight and sound or touch, taste, and smell stimuli. Most important, they serve as one kind of raw data for communication. Theorists argue whether it is the sign that communicates or our translation of it into some verbal word. Our model makes that consideration unimportant. Whether the stimulus is ordered or a random sign, it still can be sent forward to a receiver, and that receiver can make some message of it. The hiss of the Bunsen burner in the laboratory or the rumble of an auto engine are both signs, and both provide potential for message creation.

Symbols are the other sort of raw data. They are sensory stimuli too, but our communities make them very different from random stimuli. A symbol has regularity and thus predictability. These qualities are neither inherent nor accidental. Societies *decide* on what meanings its members can assume for certain symbols. They collectively give a rule structure for how something is going to be counted. There is nothing magic about the color white, for example, except that Western society has given it the symbolic meaning of innocence or purity. As a result, traditional bridal gowns are almost always white. Even today, it would be highly unlikely for a bride to have her gown and veil made in red. Red is just the wrong symbol. It may be accepted as a happy, bright-seeming color, but imagine the uproar in both families if the bride announced seriously that her gown was to be red.

You have probably heard repeatedly that "Meanings are in people; they are not in words." We agree wholeheartedly, as long as those who say so remember that the words (or any symbol) become associated with some meaning. Certainly people assigned the meaning in the first place, and people can change that meaning any time they wish; but until some change does take place, members of a speech community are entitled to expect some predictability when they use or perceive a particular symbol. These conventions allow us to think of symbols as stimuli that have order.

SYMBOLS AND LANGUAGE

Symbols combine with other symbols, and their very combination becomes rule-governed. At this point, the symbols become *language*. Language involves not only the agreements about what the symbols can be expected to mean, but also the assumption that those symbols will be structured in an agreed-upon manner. In English, for example, the placement of words in a sentence is important. It matters a great deal to English speakers if someone says, "James hit John" or "John hit James." Those of you who studied Latin know that, for Latin users, order is unimportant but word endings are critical. The rules of those two languages vary and learning any new language involves learning the rules as well as the vocabulary.

We do not mean to imply that language must involve multiple symbols in a sequence. A single symbol can function as language all by itself. Any parent will tell you that when a baby learns the word *No*, that single meaningful noise will fill many months of communication. Neither does the language community have to be large in order to come to agreements on meaning. Psychologists recognize a phenomenon known as *twinning* when twins or even siblings invent fully functional languages that are meaningful only to them. The numbers of the words or their users are not the crucial discriminating factors in what makes something a language; the rules that give it predictability and structure are.

It is hard to grasp the real magic of language. Words are no more than noises in the air or marks on paper. An A-OK gesture or International Sign language is only a disturbance in the visual spectrum. Yet think how much more these behaviors really represent. If we are ever to control the loss levels of entropy, we have to deal with it working on the raw data of stimuli in communication.

These raw data have traditionally been divided into categories of *nonverbal* and *verbal.* We shall look at both, examining how each functions as a complex source of messages.

NONVERBAL COMMUNICATION

Malandro and Barker say that nobody really knows when nonverbal communication came to be studied as a serious subject, although it was perhaps as early as 400 B.C.[2] While its beginnings may be uncertain, its current popularity is clear. We have serious studies regarding the impacts of color, the placement or absence of partitions, and the imposition of rules about clothing or hair length because these factors seem to affect worker productivity and morale.[3] Popular best sellers such as Korda's *Power! How to Get It, How to Use It,* Molloy's *Dressing for Success,* or Fast's *Body Language* helped the mass audience recognize the importance of such seemingly common things as space, eye contact, and gesture. With this recognition came an awareness of how persons and things communicate messages through channels not directly related to word content.

Our position on the many controversies surrounding nonverbal study is that a professional cannot ignore the potential message value of these symbols. Our definition of *nonverbal communication* is identical to the definition we used for *all* communication, except that the stimuli, whether random or ordered, are unrelated to content of words. In this sense, whether the gesture or look or kind of clothing or touch is thought to be symbolic language (and thus ordered) or a spontaneous sign (and thus random) is less important than the realization that these things do serve as stimuli capable of being decoded into messages.

Most of us think we know how and why words (verbals) are used by people, but few of us have thought about nonverbals. What purposes do they serve? What use do people make of their surroundings and objects? Theorists have spoken traditionally of at least five such purposes, and it is worth our while to consider them for a moment.[4]

Nonverbals	*Examples*
Emblems are nonverbals that are transferred directly into verbal words.	The thumb-out gesture of the hitchiker.

Illustrators are nonverbals, usually gestures, that accompany speech to enforce the meaning.	The two-handed movement of the fisherman that shows how l-o-o-o-n-g the fish was.
Affects are nonverbals that display a response to some inner state.	The irritated motorist kicking a car that refuses to start.
Regulators are nonverbals that signal turn-taking during conversations.	The looking away and then looking at someone to signal that it's time for that person to speak.
Adaptors are nonverbals that help a person to cope with his or her self.	The nail biter who deals with nerves by biting nails.

Each of these nonverbals is a common behavior. We all tend to use and observe them daily. Some occupations rely on emblems heavily. For example, in manufacturing plants where machinery noise makes speech difficult, workers develop sets of hand gestures that communicate simple messages quite effectively. The signals that scuba and deep-sea divers use while they are underwater are emblem examples too. When you stop to ask directions of someone, you will probably see illustrators at work when the person points one way, then another, to give you the visual picture of where you are to go. Some of us pound walls or throw things when we are angry and, in doing so, use affects. When you engage in a conversation, you use the "traffic light" of regulators to tell the other persons when you have finished and it is their turn or when you wish to change the subject. You do so by nodding your head, shifting your eye contact, raising your voice, or turning your body. Finally, all people learn to cope with personal behaviors through a series of adaptors. Maybe you chew your nails; maybe you stroke your chin or fuss with your hair. These actions are self-gratifying, and they allow you to deal with your own idiosyncratic feelings. Perhaps these examples help you imagine why nonverbal communication developed.

As you can imagine, studying nonverbals can be extremely complicated. The subject matter is usually broken down into subcategories such as touch, body types, physical objects, and so forth. The list can be long indeed. For our purpose, we have selected only six subcategories that technical professionals use most often in the corporate and social world. By focusing on these six, however, we hope your awareness of other message potentials is aroused.

Paralanguage

All of the voice overtones such as pitch, pronunciation, rate, noises between words, and volume are called *paralanguage.* While the content of the words is being decoded, other vocal stimuli are too. Oral speech cannot happen unless there is paralanguage. Think of it this way: if someone were speaking in the next room but out of your sight, could you tell if the speaker were male or female? Could you guess the approximate age of the speaker? Could you tell if the speaker were from New England or the South? Most of us could. Studies in paralanguage indicate that we make assumptions that are even more far-ranging than gender and age. In experimental settings, test subjects attribute status, intelligence, likability, and even honesty to vocal overtones.[5] If the test data are accurate, then you should consider the implications of sounding the way your listeners assume that a professional ought to sound. This book is not the place to introduce specialized vocal and articulation exercises, but many books on the subject do exist. We urge you to check your library for these sources or contact your campus speech correction clinic. Use the advice given you to improve any paralinguistic flaw. It will be well worth your time.

History of the motion picture industry is filled with sad stories of silent-screen heroes and heroines whose voices did not fit the screen image they portrayed. They were visual standards for male and female beauty for their time. When talking pictures added the star's voice to the standard, some of their careers ended. The image of an all-male swashbuckler faded when it was discovered that he had a high squeaky voice. The sultry silent vamp who projected such a fantasy of the alluring woman of the world became ludicrous when fans learned that she had a thin shrill delivery. These film disaster stories are true, and they hold a painful lesson for the effective communicator. *While* the words are being decoded, our listeners are making many judgments about us too, most of them based on nonverbals. Some of those judgments would probably shock us.

One of us knows an extremely competent engineer whose career is at a standstill. He is never allowed to meet with customers, and his appearance at hearings is always brief. Through a series of unfortunate habits, he projects an extremely affected and aloof personality. People complain that he seems to "know it all." Among the different nonverbals that communicate this sort of message, his voice patterns call attention to themselves. He speaks in long drawling tones, utilizes too much air so that his speech seems filled with sighs and has a very slow delivery. Together, these nonverbals project a person seen as cold and filled with his own importance. The picture is not accurate, but people who do not know him well interpret him according to what they see and hear. His own nonverbals, especially voice, serve as his most damaging entropy.

We observed a salesman presenting the advantages of his company's semiconductor-material slicing machine. He had speech impediments that verged on a stammer. The group of a dozen or so middle- and upper-level managers seemed sympathetic to his problem and sought to excuse it. Nevertheless, most of them were trying so hard to *avoid* noticing the speech pattern that they had trouble focusing on his message. Again, the man served as his own contribution to entropy.

Your voice and pronunciation are very personal. Many college freshmen are shocked to learn that the rest of the world does not sound like the folks back home. A young person coming from an insulated neighborhood, where everyone shares the same vocabulary and sound pattern, is sometimes ridiculed in college because, "you sound funny." That strikes at the very ego, and it is hard to accept. We do not suggest that each of you should sound like a network announcer. (These announcers, by the way, work very hard to rid their speech of any regional dialect.) What we do want you to keep in mind is that your voice itself functions as a stimulus being decoded by your receivers. If you are pleased with the sort of message those receivers get, then paralanguage is being translated accurately. For many of us, the message needs work.

Is it true that you are stuck with the voice you have? What can you do to change that with which you were born? The answer to the first question is, "Only in part," and to the second, "A great deal." Your basic vocal mechanism is a matter of biological reality, but the range available to that mechanism is broad. If you have ever heard a good vocal impersonator, you have the sense of how much you can alter your voice production. People can sound "just like" John Wayne or Burt Reynolds one moment and Clint Eastwood or Rodney Dangerfield the next. The impersonators do not have voice mechanisms like those of the people they imitate, but they have learned the possible range and alteration of their own voices. You also have choices as to how you will sound. You can learn to make the patterns differently, articulate the words in another way, slow or speed up the rate. It is really not just a matter of "sounding better." Someone may be translating your paralanguage into an inaccurate message about your personality and competence. Who can afford such entropy?

Objects and Color

Physical things and their colors are stimuli that can be translated into messages in many ways. Our worlds are filled with deep plush rugs or threadbare sweaters, crosses or stars of David, persons with long blond hair or spiked punk-rocker peaks. As receivers, we take in these objects and make message-sense of them.

For convenience, let us subdivide the subcategory of objects and color. We will consider the human body, the clothing and adornment of that body, and the surrounding environment. These message potentials seem especially pertinent for those working in corporate centers.

The shape of the human body, usually called body types, functions as an object message. That realization seems startling until you recognize how much stereotyping is based on the type of build a person has. Quickly, now, list three adjectives that describe a short, heavy man's personality. If you thought of such things as, jolly, easy-going, relaxed, content, you would be following the way our culture tends to see people who are overweight. In the 1940s, William Sheldon formulated a way of categorizing body builds:[6]

Endomorphs	*Mesomorphs*	*Ectomorphs*
Heavy, oval-shaped, soft	Muscular, trim, athletic, hard	Fragile, flat, very thin

Sheldon asserted that people possessed certain personality traits that went along with the body types categorized. Endomorphs were slow, friendly, and emotional; mesomorphs were confident, dominant, and temperamental; ectomorphs were tense, withdrawn, and fussy. Overall, people tend to prefer mesomorphs. Whether the person actually does have this type of personality or whether society *assumes* that a person does is a matter of controversy. What seems clear is that a person's body does tend to raise assumptions about that person in the minds of receivers.

What the length and style of hair should be, whether men should have beards or mustaches, and whether females should shave their legs or armpits have all been the subject of numerous debates. Endless arguments with teenagers about getting a haircut or with military recruits about facial hair seem trivial until you realize that the underlying point is sound: people do draw messages based on the distribution and length of hair. What the accuracy of that message might be is the point of controversy.

Skin color has been a persistent and highly emotional issue in the United States. That persons draw message conclusions from the simple variance of a skin tone, whether black, yellow, red, or white, seems beyond question. The morality of its importance remains a matter of human dispute.

Clothing and the adornment of the body, regardless of body type or color, are yet further considerations of object nonverbals. Perhaps it is too strong to say that, "Clothing makes the man (or woman)," but we all draw considerable messages from what a person wears and what color the clothing is. The bride in a red gown and veil shocks no less than the infant dressed in black. Imagine the impact of a man arriving at a job interview dressed in a well-tailored, three-piece, *purple* suit.

Women who work in traditionally male professional jobs complain that the rules of clothing are imposed on them with great ambiguity. The "little girl" look of ruffles and full skirts distracts from an image of competence, yet the shirt-and-tie uniform designed for males will not do either. Women, according to Angier and Sutton, took Molloy's advice and ended up looking like "diminutive men." The 1980s seem to have brought some clarification and expansion of choice. Attractive dresses, stylish colors, and women-tailored suits are being worn and accepted as appropriate even in the conservative "navy and pin-striped" banking businesses.[7]

Before the widespread use of calculators, technical professionals-in-training took great delight in wearing slide rules attached to their belts. Some of these slide rules were eighteen inches long, and their slapping against the thigh was not comfortable; but they were such an identifying mark of the engineering or physics major in the 1950s and 1960s that they were worn with pride. Calculators ended that. Since most college students have one now and some even wear them attached to the belt, they can no longer identify a person's major accurately.

Corporations, like schools, have standards of identification for their members. Wholesale rules about what everyone must wear "for success" do not exist, but any company has specific norms. They are rarely posted, and sometimes nobody tells you about them either. A new hire spends much of the first few days on the job figuring out what is or is not acceptable in dress. The classic three-piece pin-striped suit for the male or female might be wrong for a company whose code runs to more casual dress.

Some companies encourage white lab coats for the technical staff, and we know of one company where coats are worn even if the person never sets foot in a lab, because the coat itself is a sign of technical competence. Some companies use the white coat merely as a cover to keep clothing clean. A supply is kept outside the door of the test chamber, and everyone who enters the room is expected to put one on. No status at all is attached to wearing the coat. Another company we know uses two lengths of lab coats. The engineers wear the long ones, and the technicians or visitors wear short ones. It is a source of great irritation when someone who is "not entitled" to a long coat puts one on by mistake. We have never seen a lab coat in pink, by the way. That color of coat is often worn to protect the clothing of hair-dressers, yet more than simple clothing protection is at issue here.

The rules regarding lab coats or three-piece suits might seem silly to outsiders, but they do form message potentials for the company community. Each of us has to decide how we wish to be perceived, and part of that perception is based on the clothing we choose to wear.

Physical objects form an intricate system for communicating status within corporations. The office of the chairperson of the board and that of the most recent technical hire offer a fascinating set of measures to establish who is who. We usually think of an office as consisting of utilitarian furniture and equipment, and of course it does; but it tells us much more than that. William Jurma, an office communication researcher, found that executives follow very consistent patterns even in something as simple as desk layout. The desks are usually wood, supplies are in order, and telephones are dark.[8]

One company we observed furnishes management offices (and these are the only private offices in the building) with metal desks, approximately 30″ by 60″, metal chairs with padded seat and back, a file cabinet, and—when the manager has attained a certain status—a bookcase. Upper management has the identical furniture, but the desk is 36″ by 72″. Section managers always have wooden furniture. In addition to the desk and a fully upholstered high-back chair, there is a couch, a cocktail table, and a credenza. The vice president's office has, in addition to everything the section head has, several couches, a private closet, and a mini-kitchen hidden behind artistic screening. The office walls of the section managers and higher executives usually have works of art and framed pictures of projects the occupant has headed. It seems acceptable for all managers to have project pictures, but these are not framed until the section level.

Carpets seem to carry another unusually revealing object message. For the company we have been describing, no carpets appear in any office until the section-manager level. The quality and plushness of floor-covering increases as the managerial level increases. On "walnut row," a series of offices housing a number of vice presidents, the color and type of carpeting from each office extends into the hall forming a "small rug" effect that seems to announce its occupant. Whatever the carpet, it must be different from the floor covering of the hall.

The way this particular company uses office objects to tell others about the status of its members is certainly not universal. Any such formula rule would be ridiculous, but office environments obviously serve as more than "efficient work space." Careful monitoring can tell you a great deal about the company and its leadership. The environment of any communication serves as its setting and, as our model indicates, the setting is a crucial communication component.

Touch

Americans engage in very little communication by touch when compared to people in other national groups. Corporate standards seem to follow the general. One is expected to shake hands upon greeting and leaving a new acquaintance. If, however, that new acquaintance becomes a frequent visitor, the handshakes stop unless a long time has passed between meetings. The rule seems to be that new or long-time-absent encounters permit handshaking, but little touch at other times.

Men are still often confused about the propriety of handshakes when the new acquaintance is female. It is amusing to watch the subtle hand movement, withdrawal, and movement again that often accompanies such meetings. Some women solve the confusion by offering their hand first, ending the uncertainty so that the greeting can proceed. Perhaps as more women enter into the technical disciplines and find their way into corporations, such social embarrassments will cease.

What is unlikely to alter is the social conditioning that prohibits touch. When asked under what conditions he would touch a business acquaintance, a male biologist could think of none outside the ritual handshake except "maybe" that when he goes to lunch, he might step back when opening the door and lightly touch the other's shoulder saying, "After you." He doubted that even such a casual gesture would be appropriate for him if the acquaintance were female.

Even the ritual handshake among same-gender persons may be misplaced or misunderstood. A research mathematician told us that he worked with a group of people for over a year when he was asked to join their after-work tennis game. After the first evening of doubles, he told his three friends how much he enjoyed the game and shook their hands. They responded with feedback of surprise. He never shook their hands again, although he still says how much he enjoys playing tennis with them.

Eye Contact

American culture places emphasis on maintaining good eye contact. From childhood we are taught that an honest person is not afraid to look you "right in the eye." It is a small step to assume that, if someone refuses to look you right in the eye, that person must be a shifty-eyed crook. A parent often says to the children, "Look at me when I'm talking to you." Speech teachers spend a lot of class time encouraging speakers to look at the people in their audience. The message is learned quite firmly: one is expected to maintain strong eye contact or suffer negative reactions. Mark Knapp suggests that eye movements have a wide range of human message value: downward glances mean modesty; wide eyes are thought to mean fear, innocence, wonder, or honesty; eyes rolled up mean doubt or perhaps fatigue, while "cold stares" are read as aloof.[9]

Eye behavior would not be difficult then, except that our culture also teaches that staring, whether "cold" or not, is unacceptable. The time lapse between strong eye contact and staring is extremely brief, let alone ambiguous. Some of us never learn it and move through our communicative days characterized as "shifty" or as someone who is "always staring." A wide body of research suggests that sustained eye contact comes to be interpreted as hostility, challenge, or threat.[10] For a new professional, such decoding by a supervisor has serious career implications. It also becomes easy to see how business and social gaffes happen.

The avoidance of looking is also convention-bound. Even chance glancing at some things is considered inappropriate. Some rules concerning what *not* to look at appear to be centered on those things that might be misconstrued. Staring too intently at your supervisor could lead to an assumption that you pose a challenge. The growing frequency of handicapped workers in the corporate world forces the realization that handicapped people do not want their peers to ignore the disability, but neither do they want sustained gazes to tell them that the handicap is the only thing another can see.

Likewise, office courtesy requires that people look away when something personal is happening. When the boss is chewing out a secretary or two technicians are arguing about how the drawing is to be interpreted, observers are expected not to look. In many ways we behave as we did when we were children. A little boy puts his hands over his eyes so mommy will not see him. "If I can't see her, she can't see me," becomes transposed into, "If I don't look at what's happening, it's not really happening."

Eye contact seems to be used in one highly effective way: as a means of sending "turn-taking" signals to others with whom we are in conversation. If you stand back and watch a group of people in conversation, it is amazing how efficiently the flow of words moves. Usually, there is very little talking together. Instead one person finishes, another picks up the topic, and still another offers a chance comment. When two people do both start saying something at the same time, they apologize, since that is not supposed to happen. Researchers think that regulation of whose turn it is to speak is communicated largely by eye contact. Malandro and Barker summarize some of the findings by pointing out that listeners look at the speaker increasingly as their turns for speaking approach, that speakers respond by looking away and then looking directly at the next turn-taker, that speakers are less likely to break eye contact before the end of a question.[11] If you have participated in a conversation where someone seemed unaware that the right to speak was supposed to be shared, the usefulness of eye contact as regulator is a welcome tool.

If you play poker, you probably know how much of the game is bluff and how much is due to a good hand. Professional card players develop a deadpan facial expression known as the "poker face" specifically because they do not want their faces to communicate what is what. Researchers have found, however, that pupillary contraction and expansion are also very informative to those who know how to read such signs. When a pleasing subject appears, our pupils dilate; but they contract when the subject is unpleasant.[12] Perhaps for this reason, really careful poker players, and maybe labor negotiators too, keep their heads down when dealt that very good counteroffer or abysmal hand. Who knows who might be monitoring the pupils across the table?

Proxemics

Edward Hall coined the terms *personal space* and *proxemics* in his landmark study of the way different cultures utilize physical space.[13] The terms have become very familiar, and there is continued interest in studying how space differences affect behavior. For the business world, the division of work areas may have implications far beyond simple efficiency. Where are workers placed? How close is the nearest coworker? Is there somewhere that belongs *only* to a certain person? What happens when someone invades that special space? Is there a public place that everyone shares equally? Do workers alter behavior in this public space? How does status relate to allocation of space? Answers to these questions might reveal why some organizations are productive and happy places, while others suffer enormous worker dissatisfaction and turnover. In terms of communication, we see space manipulation as one possible stimulus that, along with all the others, results in a message decoding.

Coincident with the American reluctance to engage in touch is American resistance to standing close. To see Hall's personal space theory in action, it is fairly simple to demonstrate what your own or another's comfortable distance might be. Engage someone in face-to-face conversation. Experiment with moving closer or backing away. When you feel the space is about right, you have found your comfort index. The problem, however, will be with the other person to whom you speak. As you move forward to test your comfort space, your partner will probably move backward. You can literally move a person right around the room with this when-I-step-forward-you'll-step-backward ritual. Try it sometime during a dull party, but be very cautious. We tend to permit people close to us in the face-to-face situation only under certain conditions, one being the situation of intimacy and attraction. Your message may be decoded as a turn-on. What will you do when your partner stops backing away?

The corporate setting sometimes forces people into confinement. Elevators get overcrowded; waiting rooms fill unexpectedly; emergency general announcement meetings jam too many people into too little space. People deal with this situation in ways that the literature predicts. As people get off the elevator, those remaining move apart; objects such as briefcases or file folders are used to protect some of

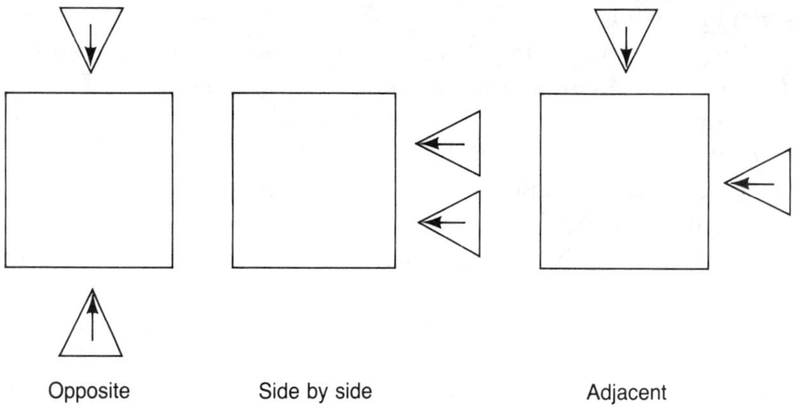

Figure 3-1 Seating positions at a table affect the probability of conflict.

the space around people in crowded rooms; during crowded meetings bodies are positioned to touch as little as possible.

Much has been written about what seating arrangements yield what sort of exchanges. By now most of us know that sitting opposite someone with a table in between makes for a greater possibility of conflict than does sitting side by side or at adjacent sides. See Figure 3-1. Recognition of such a simple solution to the enhancement of interpersonal communication holds promise and should be kept in mind. Be sensible, however. Do not expect miraculous changes in behavior because you have moved a few chairs.

Think for a moment of those offices we described above. Where do you suppose "walnut row" is in proximity to other offices? If you guessed that it is difficult to reach and well protected by the offices of secretaries or aides, you are entirely correct. A rule of thumb for American offices is that the higher the status of the occupant, the more obstacles one has to overcome to reach that office. A receptionist, usually the lowest status position, sits nearest to the door with no shielding from anyone who might come in. The person with highest status has a well-protected office. Not only is it usually far from the front entrance, but it is also surrounded by one or more offices through which visitors have to move.

A senior-level officer of a midsize company we observed claimed that he had an "open-door policy" and that anyone could drop by any time. Even a little observation showed that not too many people dropped by, unless they had an appointment. His office was on the second floor of a modernistic building. On the first floor was a reception area. The receptionist sat in a large, beautifully paneled room. Her desk was about one-quarter of the way into the room. To her left and right were doors that led to first-floor offices. No one could enter or leave those doors without her seeing. At the rear of the room, well behind her, was an open staircase. The stairs were the only way to the second floor. Again, nobody could approach the stairs without her seeing and, since the staircase was open, a person was visible all the way up.

At the top of the stairs was an open foyer, able to be seen from the first floor. The executive office was in a wing to the right of the stairway. It was glassed-in and entered through two large doors. Two clerk-typists sat in the glassed-in area. Behind them were dark wooden doors that led to the office of the executive's private secretary. To her left and never out of her sight was the door leading to the "open door" office. The door, by the way, was closed. We are not sure if the man really did wish to be accessible to his subordinates. If so, the office management of space would discourage any but the most courageous. Room and placement message was clearly, "Approach with awe. This person's time is vital. He cannot be disturbed by just anyone." How sad that the man never knew he was sending a conflicting message.

Kinesics

Many people think that nonverbal communication *means* body movement. Perhaps such an assumption is understandable because so many popular books and magazine articles have been written about how to cross your legs to signal disinterest, how to tilt your head back for assurance or how to tell if the object of your interest is "submissive."[14] Another reason may be that some of the early work done in nonverbals was by Ray Birdwhistell, a researcher whose prime interest was body movement. He named the messages people decode from muscle movements, *kinesics*.[15] You, by now, know that nonverbals encompass far more than the way a person moves some part of the body.

We cannot deny, however, the importance of kinesics in sending and decoding messages. We infer, rightly or wrongly, that the person drumming fingers on the table is nervous or that someone clenching a fist is angry. We realize that a friend is "not right today" when the body moves slowly and heavily rather than with the light sprightly movements we expect. We recognize that some gestures are taboo in polite society, and we would be greatly embarrassed if a friend made one in our grandmother's living room.

Together, the way a person disciplines the movement any set of muscles will make helps observers classify that person. That manager we described earlier with the ponderous-seeming paralanguage reinforces his seeming aloofness with an almost immobile body. His face would do justice to the poker player; he seldom gestures; his walk is maddeningly slow. He *says* he likes dealing with people. His body and other nonverbals say something else. Indeed, his problem is not uncommon. That is why speech teachers spend much time encouraging students to free up their gestures. Few of us stand normally as still and immobile as beginning speakers tend to do. The message sent is simply not accurate.

Kinesic messages in a business community require the same sort of control as does any other set of stimuli, and no formula approach is possible. Smiling is considered desirable in the business world, but a toothy smiler appears false. Facial feedback is good, unless you are dealing with a competitor and restraint is expected. Rapid purposeful walking is positive, but very rapid walking that verges on running is extremely negative. Leisurely strolling from place to place is suspect; but, if someone joins you, slowing down to a stroll is fine.

Cautions

Be careful when dealing with sets of stimuli from which messages can be translated. Whether the stimuli are verbal or nonverbal, keep the following warnings in mind to avoid serious entropy losses:

- No stimulus has meaning in itself. Those who send it and decode it give it meaning.
- The sender's meaning and the receiver's meaning are never identical, even when sender and receiver are the same person.
- Entropic loss is inherent in the differences between meanings.
- The greater the difference in meaning decoding, the greater the entropy and resultant loss.
- Language communities share some meanings for some events, but these meanings are never identical for any two members of that community.
- When a person crosses from one language community to another, the entropy increases.
- Easy generalizations about what someone else's meaning might be are filled with error possibility.

If you keep these cautions in mind about the nonverbal stimuli around you, your ability to send and receive messages will be greatly increased. One of our students said that learning about nonverbals was like having blinders removed. You just never realized all that was out there. Once you do, the world is never the same again.

VERBAL COMMUNICATION

While all those nonverbals are bombarding you, so are words. Nonverbals sometimes are sent without words, as in the emblem use; but words always have nonverbals *while* they are being sent. When marks are written on paper, the way they appear and the color of the paper form an object message too. Marshall McLuhan claims that the way the message is presented determines how the message is received.[16] Do you see any important differences in effect as you read Figure 3-2? Of course you do. The format of the type, whether the words are handwritten or machine produced, and the boldness of the print all contribute to the message. Since written words are printed *somehow*, these nonverbal messages are always present along with the content of the words. When the words are spoken, the voice carries a paralinguistic message along with the words. This seems fit, because most of our communications are a combination of the words someone uses and the nonverbals accompanying those words.

Figure 3-2 The way words look is the nonverbal part of their message.

In this section, we assume that words function as ordered stimuli. They are a part of agreed-upon and rule-governed language. Sometimes words appear singly, but usually they are linked to other words. We mean by *verbals* the content that the rules governing words encourage.

Much has been written about verbal language. The subject has fascinated and probably will continue to fascinate people as long as the species survives. The treatment of language we consider and use in this book is the work of Alfred Korzybski and his General Semantics. Korzybski, a Polish nobleman who emigrated to the United States in the 1930s, was educated as an engineer and mathematician. He thought he saw a possibility for understanding language systems in the work of science. He wrote *Science and Sanity* in 1933.[17] There followed a number of books by people who identified themselves as General Semanticists,[18] and the movement was launched. Today, his work continues through a new generation of scholars and the International Society for General Semantics, housed in San Francisco.

Korzybski's ideas seem specifically helpful to those who aspire to technical competence because he explains language from the perspective of one educated in science. We will not attempt to summarize his work, nor refute its blind spots, but rather will pick from it some few concepts to help you understand language implications.

Inevitability of Change

Some of the difficulty language systems cause stems from the fairly static quality of the words in contrast to the in-process quality of the world being described. Language is about the world, or it is about something that occurs to people living in that world. Scientists are fairly comfortable with the idea that our physical world, including the humans who live within it, is in constant flux. They understand that everything in creation alters every second that it exists. The *you* who began reading this sentence is different from the you reading this particular word. Your lungs have taken in air and exchanged it for the air your body used and altered already. Enormous numbers of cells have died within you, and others came into being. Your body chemistry fluctuated in numerous ways. Fortunately, perhaps, these changes were so unnoticeable that you look very much the same as when you started. You *look* the same but are, in a very real sense, a different person.

The problem is that our language does not reflect this ever-present change. For the language to be accurate, we would need a completely different word each instant to name that thing which is different at each instant. Consider the confusion; such language habits would be wholly impossible. Nevertheless, we do not recognize that our words cannot reflect an ever-changing world, and therefore an equally confusing error is made. We talk about people and things as if they were static.

Writing in *Etcetera*, the journal of General Semantics, J. Randall Curtis notes yet another problem in this static-language, process-world difficulty:

Language used in a certain way can give the impression that reality itself is static. While we cannot do without generalizations, classes, categories, and names, we should realize that things change; generalizations are not always dependable or useful; classifications should not become rigid.[19]

If we perceive Barbara as "stingy" because we saw her ignore the church poor box, we assume Barbara is going to continue being stingy as if she were not undergoing change. Furthermore, if Barbara's last name happens to be McLaughlin, we might compound the error by assuming that the stinginess is caused by the classification of her Scottish background.

Because it takes a long time for some things to register their changes, our faulty senses assume that nothing is happening. The piece of plastic lying in the corner of your room will remain there for many years before you can detect its change, but your inability to notice it does not mean that deterioration is not at work on the plastic. Slowly and steadily it is, but our senses cannot pick it up. Contrary to what the plastics firms would like us to believe, plastic is not eternal.

Verbal language cannot reflect accurately the ever-shifting nature of the world, so we speak of things as if they were static. We also begin to *think* of them in that way. The General Semanticists say that language ought to function in the same relationship to the world as maps do to territories they represent. If a map told you that five miles to the south there was a river, and you found no river in the place the map designated, you would throw away the map in anger and buy one that was accurate. Our static words do not represent a dynamic world very well, but we do not, nor can we, throw them out. We are stuck with our word languages. If we remind ourselves that the words are inaccurate, their harm can be lessened. If we do not, we propagate an erroneous picture of the world that even an elementary science book shows to be false. It is as if, even when we are careful with language, we are working with a defective tool. If we are not careful, that tool deludes us into pictures of a world that never existed.

Levels of Abstraction

Think about the statement we made concerning Barbara's stinginess. Do you see anything peculiar about it? Think who decided to evaluate Barbara as stingy. The *speaker* did, yet the speaker is not even in the sentence. The sentence reads, "Barbara *is* stingy," as if that quality is part of *her,* rather than of the *speaker's* evaluation of something she did. Our sentences say things such as, "The sky is blue" or "The painting is beautiful," as if the blue were inherent in the sky and the beauty were part of the painting. To speak accurately, the sky is *capable* of being seen by *someone* as blue, and that *someone* might evaluate the painting as beautiful. Korzybski claimed that the absence of the speaker (observer) in sentences led people to forget that the sentence was actually about that speaker's process in reacting to skies and pictures.

In many ways, his point was similar to Einstein's. The relativity theory was much more than a scientific revelation. In a real sense, it was a landmark that changed the way people conceived of their relationship to the physical world. Stuart Chase says that, "In the long run, relativity may prove a more important factor in language than in physics."[20] That statement is not meant to dishonor one of the giants of scientific history. Quite the opposite is meant, for the scientific community itself had to reconstruct much of what it had believed. Absolutes such as time and space needed to be reconceived as specific steps in a specific operation. Basically, Einstein argued that *someone* had to make a measuring tool and that *someone* could not be ignored as part of the experimental steps (called operations) that defined the concept under experiment. Time absolutes disappeared to be replaced by consideration of a *someone* at some given place going through a set of given operations with a given measuring tool, all of which define the variable concept of time.

General Semanticists gave the name *abstractor* to that observer, and defined *abstraction* as the process (the operation) of leaving out details. There is always a person who takes in something such as light waves from the sky, processes that stimulus through the nervous system, selects some words to talk about it, and calls the sky blue. What results, when the abstractor gets to the words, is the sentence; but so many details are left out. The words are obviously not the sky nor the sense experience of processing light waves. The process (operation) started with a physical perception and ended up with noises spoken in air or marks made on paper.

Korzybski's science training taught him that when people were working as scientists, their language tended to be extremely precise. The scientist reporting on some experiment is usually very careful to identify who made the observations, under what conditions they were made, and what exact results were obtained. Korzybski urged that some of this precision ought to be used in daily language too. The closer the words are to what *someone* perceived, the lower the *level of abstraction*. The more the words move away from the details of the observation, the higher is the *level of abstraction*. Low levels of abstraction allow for precision, while high levels of abstractions do not.

Korzybski's successors called the pictorial model of moving from low to high abstraction, *the abstraction ladder*. The analogy is a handy one, for it reminds us that the higher we move on the ladder, the farther away we are from the actuality of the observed event. Suppose I want to tell you about something in the room with me right now. I could call it all of the following, but at which level do the words help you form the message of what I want you to see?

It

Possession

Living Thing

Animal

House Pet

Dog

Cocker Spaniel

Cocker Spaniel, responds to "Shraebee," weighs twenty-five pounds, appears to have beige, buff, and brown-colored fur, dark-brown freckle-like markings on the right side of beige-colored muzzle.

If you started at the top of the ladder, you probably could not figure out what was being talked about until about the "dog" level, and it would not be until the description that you might pick out Shraebee from a room full of dogs. The higher you are on the ladder, the more trouble you would have with a specific identification of anything a sender wanted you to picture.

Place yourself in a testing laboratory where a new piece of equipment is under analysis. What level of abstraction helps you visualize the item best?

Thing

Machine

Engine

Energy Conversion Unit

Power Generator

Thermoelectric auxiliary power unit (APU)

Hollow rectangle made of painted steel and measuring 18″ × 10″ × 8″, housing a gasoline fuel burner, a plastic gasoline tank, a semiconductor control system, stainless steel heat exchanger, 300 thermoelectric couples, one fan, one pump. Painted in red on the rectangle is "APU-1."

In a room full of equipment, which words would help you retrieve APU-1? Since the object is unfamiliar, it might not have been until the lowest level of abstraction.

Suppose a Coast Guard representative came to your ship design firm and said:

We want you to design us a new ship. It has to be the latest weapon in our fight against crime on the high seas. It needs to be tough and able to house a necessary crew.

About all you could decode from this language is that the Coast Guard is in the market for a new ship design. The representative is operating at too high a level of abstraction. The staff would have to spend many hours and successive drawings moving the speaker down the abstraction ladder to what *precisely* was sought. The fundamental aim of science is to explain some phenomenon as precisely as possible.

Here is a lower level of abstraction, taken from the *Journal of the American Defense Preparedness Association,* which gives a much clearer picture of the desired ship design than does the representative's description noted earlier:

The vessel was to have a maximum speed of 30 knots.

It had to be able to attain a 25-knot speed in seas of 10-foot significant wave height.

It was necessary that the cutter be able to operate over an area with a 500-mile radius at a speed of 25 knots.

The ship had to be capable of missions of 7 to 10 days with underway refueling.

There had to be accommodations for 18, which was to include a crew of 14 and a 4-man boarding party.[21]

The designers finally have enough details as to what the representative wants for the Coast Guard. Think of the hours of talking, thousands of drawings, and volumes of calculations before the drawings are ready for approval. At that point, perhaps the representative will say, "Yes, that's it. That's what I had in mind." Think also how unlikely it would have been to get to the drawings at all, let alone the ship, if the only words were those that talked about a tough "latest weapon."

Jargon is a special problem in abstraction. Many investigators have become concerned by its use in disciplines and what jargon does to those outside of the inner group. Staff reporter Al Haas, writing for *The Philadelphia Inquirer,* asks, "why do people in specialized fields such as computers inflict their jargon on us laymen? Even worse, why do we outsiders pick up so many of these words and put them into general use?"[22] Haas looks to several linguists for an answer, and they suggest that

- Jargon gets information across quickly to those within the field.
- Jargon identifies members of a given field to each other.

Such activities should mean efficient and clear communication, admirable professional goals. Jargon, however, has other uses too, and these have serious negative communication effects:

- Jargon is used to impress the layperson by using unfamiliar terms.
- Jargon is used to demean the layperson and so create a sense of inferiority.
- Jargon is used to make the listener feel left out.
- Jargon is used as an act of hostility to diminish the listener while making the speaker feel superior.

William Labov, professor of Linguistics at the University of Pennsylvania, is quoted as saying, "I'm not kidding when I tell students that learning the right jargon is the first step to becoming a professional." Labov warns that it does cause serious communication problems. It is heartening to note his recognition that the best minds in the fields are usually those who can put their ideas into simple everyday language.

Jargon, like any other language system, can be used with those who understand it well as a low-level abstraction capable of encouraging clear and concise meanings. Used to address people who have no knowledge of a discipline or used in a filled-with-popular-error manner, it can be the source of serious entropy. Technical persons must be on guard to understand the type of listener doing the translations. We shall have a lot to say later about adapting a message to an audience. Many technically trained people forget that not everyone is familiar with the jargon of the scientific world. When messages have to be decoded by someone who does not understand a particular term, the sender has to move to a lower level of abstraction.

Fact/Inference

The foundation of science is that all ideas must be checked in the physical world. It is fine to assume that Process A or Process B will work, but the necessary next step is to subject A and B to a rigorous test observation. What then? Can the conclusions reached be called *truth?* Hardly. Truth is fixed and complete; scientific conclusions can never be. Einstein used the phrase "new realities," created by the advancement of physics, to describe the ever-changing pictures of the world scientists develop.[23]

Einstein's philosophic discussion is extremely important, because the distinction between unchanging truth and scientific "new realities" defines the *scientific method.* An experimenter, describing the results of a careful test, can claim that, *based on these data and with these operations, the following results were reached.* Such results are fact but never absolute truth. For example, each time Louise drops something from a higher plane, it falls downward. She has been told that the phenomenon is due to gravity. She tests this downward motion in a laboratory under carefully controlled conditions and, sure enough, gets readings on an electronic stopwatch that registers drop time between two points. "Hmm," Louise concludes, "that's gravity at work. The readings aren't exactly what the formula says, but I guess that's because ideal conditions don't exist here." Now what can she say factually? She can only describe the operations she went through and what result she reached. Nothing more. It is fact that Louise used a certain apparatus, that she collected her data in a specified way. It is also fact that she read certain values on her measurement tool. It is *not* fact that gravity was responsible. It *might* have been gravity. Her procedures and the approval of the scientific community make gravity a highly likely guess. Since gravity is not observable directly and its presence goes beyond anything Louise did see, it is an *inference.*

According to Korzybskian General Semantics, *factual statement* language about a perceived event is only possible after the speaker did the perceiving and is restricted to what was perceived. Language that does not meet these criteria is *inferential.* Please notice that neither type of statement is better or worse than the other. People do and should use both. Note also that neither statement claims unconditional truth. Perhaps that is the point: the *condition* of the observer is necessary for there to be fact.

Suppose the following:

You are standing at the intersection of Forrest and Main Streets. You notice a person in a black coat walking toward you on Main. The person is carrying a grocery bag with the name Acme Market printed on its side. Just before crossing Forrest the person puts the bag in a litter can on the corner. The person goes into a building farther down the block, out of your sight. A moment later you see a person in a red jacket exit the same building, walk up to the litter can, remove a grocery bag with the name Acme Market, and move quickly down Main out of your sight.

What could you say factually about this event?

- That you saw a person in a black coat carrying a grocery bag marked Acme Market.
- That you saw this person place the bag into a litter can and walk into a building.
- That you saw a person in a red jacket walk out of a building, go to the litter can, and remove a grocery bag marked Acme Market.
- That you saw the person in a red jacket walk down Main out of your sight.

Each of these is something you yourself perceived, and you have stated only what you perceived.

Why is it inferential to say:

- The event was a drug drop and pickup?
- The two persons knew each other?
- The same Acme Market bag was placed in the litter can and retrieved?
- There were two different people involved?
- The person in the black coat and/or the red jacket knew what was in the bag?

Each of these inferences might be accurate, but they go beyond the observer's perceptions. They are guesses.

It seems obvious that much of what we do and conclude daily is based on inference. Nothing is at all wrong with this behavior, so long as we are willing to remember that a guess is only probable and might later turn out to be erroneous. Think of all the terrible accidents that happen because, "I was just sure that there was no car coming on that one-way street" or "The light was red for him. I was just sure that he was going to stop" or "I was just sure the gun wasn't loaded."

Facts can be in error too. Remember, we are claiming that they are not *truth;* they are only what someone perceived. Our senses, after all, are not very sophisticated measuring instruments. You could have seen a black coat. That is fact for you, but the coat could have appeared dark blue to a hundred other people. There was a sensational murder trial in Philadelphia a few years ago. A police officer was on trial for his life in the shooting of another man. The shooting took place in a small diner with at least fourteen eyewitnesses on the scene. Some saw the murdered man with a knife; some saw him put money into a pay phone and talk to someone; some saw him push the officer; some saw the officer push the man. All of them were speaking factually, reporting only what they perceived, and no two witnesses agreed on what happened. The officer was acquitted, because the jury found the testimony confusing enough to prohibit a conclusion "beyond a reasonable doubt."

Probably the most difficult thing about separating fact from inference is the disciplining of language so it reports only what was actually perceived. Our minds want to jump from an observation to a conclusion because "it just makes sense." It may, but the conclusion is inferential nevertheless.

Take this test. Read the paragraphs as many times as you wish, then answer the questions as *fact* if that is what the paragraphs say and as *inference* if the question forces you beyond what the paragraphs say. The paragraphs do not "lie." If they say that there was a woman who closed the door, it was a woman rather than a man dressed as a woman. (Your daily life, of course, is filled with such "lies.")

Jackie Wade, a sophomore computer-science major, was confused about one of the college's general education requirements. Jackie took the question to Professor Gallen, Chairperson of Advisors for Computer Science. Professor Gallen was not in the office at 2:30, even though the posted schedule clearly states that 2:30–4:00 P.M. were Gallen's office hours. At 4:00 P.M. Gallen arrived at the office and noticed Wade standing in the hall. "I'm really in a hurry. Is there something I can do for you quickly?"

Jackie's face was flushed. "I think so. If it wouldn't be too much trouble, I have a question about this statement in the catalog. It says that freshmen are to take 100-level courses, sophomores take 200-level courses, but juniors or seniors can take 300- or 400-level courses. Is that accurate?"

Picking up a briefcase and putting papers inside, Gallen said, "Obviously since that's what it says, that's exactly what it means."

	Fact	Inference
1. Professor Gallen is Jackie Wade's advisor.	x	x

	Fact	Inference
2. Wade was upset at how the general education requirements would affect him.	x	x
3. Professor Gallen's office hours were posted in the office clearly.	x	x
4. Professor Gallen is careless about keeping office hours.	x	x
5. Gallen did not want to talk to Wade just then.	x	x
6. Wade waited for Gallen an hour and a half in the office hall.	x	x
7. Jackie realized Gallen was in a hurry and answered with annoyance, "If that wouldn't be too much trouble."	x	x
8. Jackie Wade, a sophomore, has no 300- or 400-level classes.	x	x
9. Sophomores cannot get credit for any 300- or 400-level courses.	x	x
10. Putting papers inside his briefcase, Gallen told Wade that the catalog means what it says.	x	x

Each of these statements might be accurate, but all are inferences. Look at them again:

1. Gallen is Chairperson of Advisors but not necessarily Jackie's advisor.
2. You don't know how Wade felt, only that "Wade's face was flushed." Fever? Embarrassment? You also do not know why Wade is interested in the catalog statement. Who said that Wade was male?
3. You do not know where the office hours are posted.
4. This could have been the first time Gallen ever missed an office hour.
5. Gallen might have wanted to talk to Wade but could have been unable to for reasons beyond personal control.
6. You do not know where Wade was during that hour and a half.

7. You do not know that Wade was annoyed. Neither do you know if the statement was said sarcastically.
8. You have no idea what classes, if any, Jackie has.
9. Who says so?
10. What makes you think Gallen is male? Do you know who owns the briefcase?

Maddening, isn't it? Our minds want to jump to some of these conclusions. They made *sense,* and some or all of them might be accurate, but some or all of them might be completely wrong too. Harry Weinberg claims that knowing the difference between fact and inference makes one more cautious about conclusions and more likely to take appropriate action. He says:

If we know that we are making an inference and not a factual statement, then: (1) We are less sure of what we assert, and therefore are less likely to assert it dogmatically. (2) We are prepared to be proved wrong, and therefore are much less likely to be hurt, physically and emotionally, if we are proved wrong. (3) We expect the unexpected and therefore are more prepared for it.[24]

The statement could serve as the corollary to the scientific method. *Dogmatic scientist* ought to be a contradiction in terms. Some awareness of the way language tricks us into assuming more than we perceived about an event is a valuable disciplinary exercise for technical professionals.

Many scientists perform their professional tasks with extreme care, and they use language precisely when they are in the testing laboratory. When they hang up their lab coats, however, the care is turned off with the test-cell lights. One of us observed a highly respected industrial chemist while he was finishing a test on a bond material. His language was precise. Rather than saying, "That looks better than the X-9 stuff," he lowered his level of abstraction with statements such as, "The aluminum-filled epoxy lowered the temperature differential at the joint by a factor of five" and "The aluminum-filled epoxy costs one-tenth the price of what silver-filled cost." His written report was fully documented, quantified, and defined. Yet this same man opened the door of the test chamber and shouted down the hall to his secretary, "Bring me that file on my desk." When she brought him a stack of files, all from his desk, he was annoyed because she did not know which one he meant. It is very hard to excuse the lack of verbal awareness in one who demonstrates that he can use language so well.

WRAP-UP

Are we suggesting that persons walk around quantifying and defining everything they say? Not at all. As with every part of the communication process, language is filled with entropic losses. Sometimes it is right and proper to be obscure. When your friend shows up wearing a wild Hawaiian print shirt in purples and hot pink, you might not want to say, "Take that off. The last three people who walked by stared at you and smirked." Instead, protecting your friend's ego, you say, "That's . . . interesting."

Some of our most beautiful poetry is extremely abstract. Perhaps that is why so many different people find so much in it. There is a place for poetry, just as there is a place for the scientific report. The difficulty comes when a speaker does not know which is which. Some of the entropy in verbal stimuli might be controlled if people remembered that languages are inherently faulty tools. They represent the world as static, when it is demonstrably dynamic. Users of language forget that words are not born equal. Some involve extremely high levels of abstraction, while others communicate with low-level loss potential. Some jargon, for example, can be either precise or totally confusing, depending on the listener's background.

Extremely dangerous for the technical professional are the hidden traps in talking about the world, especially statements of fact and statements of inference. Being conscious of possible problems with the verbal language components and working to keep them under control can cut down on entropy.

Equally dangerous is the ignorance many of us have regarding the message-carrying potential of everyday nonverbals. Whether for social or career success, factors of paralanguage, objects and color, touch, eye contact, space, and body movement are important complementary language systems that accompany the verbals.

We listed a number of cautions at the conclusion of the nonverbal section. It is now time to remind you that *both* verbal and nonverbal stimuli come at you fairly constantly. Likewise, you send them out persistently. Remember the cautions. When communication goes astray, as it always does *in part*, look to the random and/or ordered verbal and/or nonverbal components for some understanding of how the message went astray.

ACTIVITIES

1. Communicate to the class some real or imagined problem that you have and the solution you might use to solve the problem. Use any symbol system you wish, but no written or oral verbal language can be used.
2. Each person in class should present a 3–4 minute presentation on some current issue highlighted in the news. *Ignoring* the content of the *words*, each member of the audience should focus on the speaker's nonverbals during the presentation. How could you tell if:

   - The speaker felt at ease with the topic?
   - The speaker believed the statements made?
   - The speaker is someone you might want as a friend?
   - The speaker is someone you would trust?

   What do your conclusions about the speaker tell you about how nonverbals influence impressions?
3. Keep a log of inferences you made over a weekend and thought were accurate facts. How many of these turned out to be other than what you thought they were? In 2–3 minutes, report to the class how something you thought was accurate ended up being in error. Did any of these cause you problems or embarrassments?
4. Check the newspapers for accidents or errors that resulted in misunderstandings. What inferences did those involved in these events make? What kind of checking could have been made to avoid the results reported?
5. Select a paragraph from some science textbook you are using in another class. Still keeping to the basic content of the paragraph:

   - Rewrite the paragraph at a lower level of abstraction so it is clearer than its current level.
   - Rewrite the paragraph at a higher level of abstraction so it is more abstract and perhaps more confusing.

   What problems do the first and second rewrites cause?

NOTES

1. For a full discussion of these controversial terms, see F. E. X. Dance, ed., *Human Communication Theory* (New York: Holt, Rinehart & Winston, 1967) or J. Burgoon and T. Saine, *The Unspoken Dialogue* (Dallas: Houghton Mifflin, 1978).

2. Loretta A. Malandro and Larry L. Barker, *Nonverbal Communication* (Reading, Mass.: Addison-Wesley, 1983), 4.

3. See, for example, H. Ketcham, *Color Planning for Business and Industry* (New York: Harper and Brothers, 1958); "The Trouble with Open Offices," *Business Week* (August 6, 1978), 84–88; L. B. Rosenfeld and T. G. Plax, "Clothing as Communication," *Journal of Communication* 27 (1977): 24–31; and Alison Lurie, *The Language of Clothes* (New York: Random House, 1981).

4. P. Ekman and W. V. Friesen, "The Repertoire of Nonverbal Behavior: Categories, Origins, Usage, and Coding," *Semiotica* 1 (1969): 49–98.

5. See, for example, F. Williams, "The Psychological Correlates of Speech Characteristics: On Sounding 'Disadvantaged,'" *Journal of Speech and Hearing Research* 13 (1970): 472–88, or E. Kramer, "Personality Stereotypes in Voice: A Reconsideration of the Data," *Journal of Social Psychology* 62 (1964): 247–51.

6. William H. Sheldon, *The Varieties of Human Physique* (New York: Harper & Row, 1940), William H. Sheldon, *The Varieties of Temperament* (New York: Harper & Row, 1942), and William H. Sheldon, *The Atlas of Man: A Guide for Somatyping the Adult Male at All Ages* (New York: Harper & Row, 1954).

7. Natalie Angier and Tina Sutton, "Dressing Up the Organization," *Savvy* (May 1984), 32–36.

8. "Executive desks: Wood, and orderly," Philadelphia *Inquirer*, 14 January 1984, Sec. C.

9. Mark Knapp, *Nonverbal Communication in Human Interaction*, 2nd ed. (New York: Holt, Rinehart & Winston, 1978), 294–95.

10. Albert Mehrabian, *Silent Messages*, 2nd ed. (Belmont, Calif.: Wadsworth, 1981), 51–52.

11. Malandro and Barker, *Nonverbal*, 164.

12. M. P. Janisse and W. S. Peavler, "Pupillary Research Today: Emotion in the Eye," *Psychology Today*, no. 7 (1974), 60–63.

13. Edward T. Hall, *The Silent Language* (Garden City, N.Y.: Doubleday, 1959).

14. For a representative popular treatment, see David B. Govens, *Love Signals* (New York: Crown, 1983).

15. Ray L. Birdwhistell, *Kinesics and Context* (Philadelphia: Univ. of Pennsylvania Press, 1970).

16. For an explanation of McLuhan's theory, see Marshall McLuhan, *The Gutenberg Galaxy* (Toronto: Univ. of Toronto Press, 1962) and *Understanding Media* (New York: McGraw-Hill, 1962). For graphic examples of his theory, see *Counterblast* (New York: Harcourt, Brace and World, 1969).

17. Alfred Korzybski, *Science and Sanity* (Lakeville, Conn.: Institute for General Semantics, 1933).

18. Among these are: Stuart Chase, *The Tyranny of Words* (New York: Harcourt Brace Jovanovich, 1966; Irving Lee, *Language Habits in Human Affairs* (New York: Harper and Brothers, 1954); S. I. Hayakawa, *Language in Thought and Action*, 4th ed. (New York: Harcourt Brace Jovanovich, 1978); and Harry L. Weinberg, *Levels of Knowing and Existence* (New York: Harper & Row, 1959).

19. J. Randall Curtis, "General Semantics as a Tool to Investigate Scientific Empiricism," *Etcetera* 40 (Winter 1983): 410.

20. Chase, *Tyranny*, 117.

21. Paul Seidenman, "SES Joins the Search," *National Defense* LXVIII, 393 (December 1983): 44–46.

22. Al Haas, "A New Crop of Jargon Hampers Interpersonal Interfacing," *The Philadelphia Inquirer*, 2 April 1984, Sec. E. Used with permission of *The Philadelphia Inquirer*.

23. Albert Einstein and Leopold Infeld, *The Evolution of Physics* (New York: Clarion, 1938), 294–96.

24. Weinberg, *Levels*, 20.

CHAPTER FOUR

LISTENING

In Douglas Adams's *The Hitchhiker's Guide to the Galaxy*, the following conversation takes place between Arthur, the Earth man, and Ford, an alien:

"You know," said Arthur, "it's at times like this, when I'm trapped in a Vogon airlock with a man from Betelgeuse, and about to die of asphyxiation in deep space, that I really wish I'd listened to what my mother told me when I was young."

"Why, what did she tell you?"

"I don't know, I didn't listen."[1]

That story has more than a little bite to it, for at times most of us suspect that "nobody's listening." Unfortunately, we are very nearly correct. Despite studies verifying time and time again that people spend *most* of their communicative days in an activity they call listening, they tend to do it poorly. It is time to direct our attention to the decoding side of the communication model.

Perhaps you recall the Rankin experiment cited in Chapter 1.[2] Paul Rankin concluded that listening and speaking occupy over 70 percent of a person's communicative time, some 41 percent of it listening. Nichols and Stevens set the listening figure higher, at 45 percent.[3] Despite those statistics, people are not trained to listen well. A recent book on listening points out that Rankin himself was aware of how little training there was.[4] He discovered, as Director of Research and Adjustment for the Detroit public schools, that reading (the second-least-used communication skill) received 52 percent of the emphasis in classroom instruction, while the most-used skill of listening received only 8 percent of the emphasis.[5] The figures have changed little in nearly forty years.[6] We begin to have a clue as to why "nobody's listening."

Before you conclude erroneously that reading and writing (the least-used communicative skill) are unimportant, be assured that we are not saying so at all. Each of the communication skills is part of human behavior, and all should be developed to the highest possible level. What we find so objectionable is the almost total lack of attention given to the skill used most.

Does it matter? Perhaps listening needs no training and most people actually listen quite well. Your own experience, however, probably tells you that a lot of people adopt the Galactic Hitchhiker's habit of tuning others out. Patients claim that their doctors pay no attention to anything they say. Ironically, the doctors say the same thing about the patients. Teachers complain about how poorly students listen. Those same students counter with, "Did you ever try to tell *that one* anything?" Marriage counselors try to get couples to listen to each other, and sometimes the spouses of those counselors would like the same consideration.

Listening problems cause enormous difficulty in the business world. One of us conducts workshops for midsize companies, and the most frequent request for help is in the improvement of corporate listening. Raymond Ross quotes the comment of J. Paul Lyet, chairman of Sperry Corporation, that poor listening is one of the most important problems facing business. The price tag for this poor listening was computed to be in the billions for American businesses.[7]

That billion-dollar error is not difficult to understand. Ralph Nichols postulates that the average white-collar worker operates at about 25 percent of potential listening efficiency.[8] Imagine all the work that has to be redone because a secretary thought the boss said two originals and three copies, instead of the reverse, or because the parts supervisor thought the technician said Item 6-Z rather than 16-Z. Imagine the time lost by thousands of motorists trying to find their way according to directions they thought they heard. Even without figuring the emotional and interpersonal costs to these people, the monetary losses are staggering.

Although everyone agrees that listening is a real problem, it continues to be unchecked. We hope part of the solution will begin with you. When new business and technical leaders have the necessary training (which this book and your course are intended to provide), then a solution is possible. In this chapter, we will examine problems involved in listening and suggest some ways to improve your decoding.

Perhaps some of the data cited above disturb you. How can

statistics show that, on the one hand, a large part of every day is spent in listening, yet, on the other, that "nobody's listening"? Interesting analysis. Might it be that some definitions and distinctions are in order?

The term *listening*, like many other common terms, has a popular but inaccurate usage. When you were fourteen and told your dad that it would be a good idea if he bought a new Corvette so you could drive it when you were sixteen, he probably just kept buying Plymouth sedans. "He never *listens*," you sulked in disgust. When your computer-science professor assigned the deadline for the program's submission as the day after Thanksgiving break despite your loud complaints, you muttered, "I told you she wouldn't *listen*." In neither of these cases were you talking about listening, but rather about *agreement*. Your father's and your teacher's listening might have been extremely effective; but their conclusions, based on what they thought about your message, displeased you.

We might also say that the statistics are actually measuring *hearing*, rather than *listening*. People *hear* stimuli persistently, but they *listen* badly. The two functions are decidedly not the same. *Hearing* is a sensory intake, sound waves impinging on a receiving unit. In many ways it is largely passive. *Listening*, on the other hand, is an active process because it takes the impinging stimulus and consciously acts on it. The listening person reacts to what has entered the nervous system. Our communication model (Figure 2-11) calls this reaction the *translation of the stimulus*. The ordered or random stimuli that reach the receiver produce the sensation of hearing; the translation that the receiver makes is listening. How well the decoding worked for the receiver is the comparison of the *message received* with the *message sent*. In the lower part of the model, remember, the receiver reacts to the message as received and transmits ordered and/or random feedback stimuli back to the original sender. The sender *translates the feedback* as a listening function. To the degree the *feedback received* matches the *feedback sent*, the sender's listening was effective. Can you see that there are multiple functions performed by a listening receiver in every communication? In its simplest form, the receiver listens to the sender; and the sender, in turn, listens to the receiver's feedback.

A piece is still missing, however. Remember Wendell Johnson's point about the "most enchanted listener"?[9] While the sender and receiver are decoding each other's messages, the sender and receiver are listening to themselves as well.

LISTENING TO YOURSELF: INTERNAL FEEDBACK

Johnson thought that to ignore speakers listening to what they themselves have just said was to lose valuable information:

We had not often thought of speakers as their own listeners, and so we had not attended to them in a spirit of eavesdropping, as though listening in while they were talking to themselves. And now, in the spirit of eavesdropping, we can hardly help noticing that people talking about themselves and their private desperations are saying the most fantastic things to us, to themselves, that is.[10]

Once recognized as a possible communication pattern, however, these "fantastic things" can be decoded and used by those around the speakers as well as by the speakers themselves.

The way people react to their own stimuli is called *internal feedback*. We listen to ourselves constantly, taking note of what we say and do, criticizing the results, and altering future messages. This process provides an extremely valuable self-monitoring system. Picture yourself with a friend who asks, "Why do you mathematicians think everyone has to know math?" Your internal and external dialogue might go like this:

Here we go again. The last time I tried this one it took three hours. "*Come on, Al. We have to learn a lot of junk. Why not something practical?*" That was dumb. What do I care if Al thinks it's practical. "*I mean, well . . . , like . . .*" Put your brain in gear, kid. "*Look, I had to take a course in poetry that didn't exactly thrill me, but I figure: what the hell, maybe I'll learn something different.*" Oops, blew it. I forgot Al was a Lit major. "*Actually, I sort of enjoyed the class. Maybe you'll feel the same about math.*" Not bad. Hope he doesn't get all out of joint about this.

While Al was listening to you, you were listening to you also. Your reactions to what you were sending out ideally helped structure messages that kept Al "in joint."

Sometimes beginning speakers become *so* enchanted with their own messages that the internal feedback serves as a serious distraction. One of us remembers a student who was trying to deliver a speech in class. She kept stopping and criticizing her own voice. She apologized for being a bit hoarse, commented on the tremor in her words, told us how she had trouble going at a reasonable rate, and complained about not being able to pronounce a few names. Obviously, her internal listening was acute. She was following her output very well. Unfortunately, she did not seem to grasp that her dialogue should be a private affair between her own sender and receiver functions. After a few minutes, most of her listeners were paying attention only to the way her voice sounded. Not much of her message had a chance of getting through. The entropy source in that speech was easy to identify.

LISTENING TO OTHERS: EXTERNAL FEEDBACK

When the source of the feedback comes from outside, it is said to be *external.* When a father smiles, a friend frowns, an acquaintance shrugs her shoulder, or an audience roars with laughter, they are reacting to something a sender transmitted. These overt behaviors can be read by a thoughtful sender as measurements of how the message was decoded.

The communication process has marvelous monitoring and self-corrective possibilities. When the sender and receiver are two different persons, each can direct and alter what the other says and does. You probably learned that lesson long before you stepped into a classroom. Perceptive three-year-olds seem to know when it is not the time to ask Mom for a pretzel, and they wait for a more promising moment. By the time children are very far along in school, they sense quite accurately when it might be possible to stretch a teacher's rule and when it would be disastrous to do so. These examples show senders and receivers acting out and reacting to messages. Each, in part, determines what the other will say and do.

People who translate such feedback stimuli accurately earn the reputation for being "sensitive" to others. They understand what the other is saying to them and therefore are able to understand what others are feeling. These sensitive people probably also have the reputation for being "easy to get along with." They know when to back off, when to question, when to remain silent, and when to speak up. Most of us would like to be known as such a person. One way to develop these adaptive skills is to remain open to the feedback your communication partners send.

LISTENING EFFECTIVELY

The concepts of feedback and listening are inherently related. These, in turn, are tied to the entire communication process. When we talk about the different parts of the model in this book, they seem to be separate from each other. Any static model representing a dynamic system is subject to such limitations. But the parts are not separate. They overlap to a great degree, and the processes happen in microseconds.

The receiver functions are a case in point. Stimuli are sent, received (heard/seen), translated (listened to), reacted to (feedback sent), received (heard/seen), translated (listened to), and reacted to (subsequent message altered). All these functions take place while both parties are monitoring themselves internally. Since it all happens virtually simultaneously, a lot of entropy is at work.

Nine Techniques for Good Listening

No quick and easy answers to potential listening loss exist. We refuse to mislead you into thinking that some simple exercises can help improve listening. We can suggest at least nine techniques that work to keep the entropy under control. Using them could help you to listen effectively.

1. Understand the Process Everything is difficult when you do not know its system. The aim of education, as well as science, is to explain how and why things work. Listening problems are simpler to solve when you understand the listening process. When you learn the difference between hearing and listening, you can explain why you turn on your radio each morning to get the weather forecast, "listen" for a long time, and realize that you have actually missed the chance-of-rain message, although it was repeated many times. Nothing was wrong with your hearing when the newscaster spoke the words, but there was a lot wrong with your listening.

At the conclusion of the first day of a Listening Workshop conducted for a data-processing firm, one of the managers commented, "Now that I see how it works, I can never blame my wife for not telling me something, can I?" With understanding comes the possibility of making changes.

2. Compensate for Physical Problems At least four purely physical factors affect the listening process. Any of them serves as a major source of entropy.

First, you can listen almost three times faster than most people speak. Larry Barker called this difference "a lag" and pointed to some of the problems that result when there is that much free time in your head.[11] Anita Taylor et al. note that most of what you hear can be decoded at approximately 400 words per minute, while most people speak at a rate of less than 150 words per minute. Effective listeners use that time well, while poor listeners do not.[12] You are, in a sense, simply overqualified for the listening process. A chemist might be thinking about five different formulas in the lag between the time the marketing manager says that someone from the laboratory has to come to the presentation to answer questions, and the time that the manager says that the presentation is on Saturday morning.

Second, there are personal hearing problems of a purely physical nature. These might range from the temporary inconvenience of sitting near a noisy air conditioner to a major malfunction in a hearing aid. The temporary noise is, obviously, corrected more easily than is a chronic hearing disability. People can move their seats, notify the speaker that the microphone is not working, or request that a noisy piece of equipment should be turned off. Differing degrees of deafness are not nearly so simple. People suffering from a hearing disability learn to adjust to aids, lip reading, and sign language. Those of us with temporary hearing losses due to allergies or stuffy noses often do not.

Third, physical stimuli that act as sight and sound distractions are sources of entropy that are especially difficult to deal with. Anyone who has tried to converse while a baby cried nearby has dealt with sound distraction. Campuses and corporations are filled with sounds, let alone sights, that claim your attention. A supervisor we know tells employees that anyone looking out the window at the boss's Ferrari will be hopelessly behind expected output.

Fourth, physical feelings affect your listening. The human body is constantly undergoing change. Those changes make it hard for us to recognize how a small nagging headache or a faintly sore muscle gets in the way of listening. Most of us know enough to go home when we are feverish and "really sick." Most of us do not know how debilitating to the decoding function are such common states as lack of sleep, hunger, minor pains, or uncomfortable positions. These simple entropy sources probably get in the way of messages more often than most of us admit. If you do not recognize and compensate for them, these and other physical problems jeopardize listening effectiveness.

3. Listen with All Your Senses As we have used the term, *listening* means far more than simply taking in aural stimuli. All of the nonverbals come into play when you listen. You knew your brother was not going to let you use his car because he looked and sounded out-of-sorts. His words may have been extremely calm ("Susannah, there's no gas in my car"), but the way he said it and the way his face looked told you that you really had better plan on walking.

Try saying the following by stressing the word in italics:

- *Bring* that here.
- Bring *that* here.
- Bring that *here*.

The translation given to each version certainly changes. The words are the same, but the paralinguistic nonverbal altered the message potential. Ignoring such input makes accurate decoding difficult.

4. Realize That the Reasons for Listening Differ Not all listening has to be performed with the same efficiency. Some communications simply do not warrant it. Assuming that each communication is exactly like every other one leads to serious waste of effort. There are times to listen for *entertainment*. You can miss a great deal of a comedian's monologue or a TV sitcom, and it really matters very little. The purpose of the communication, at least from your point of view, is to relax and have fun. At other times, the purpose of listening is for *human understanding*. People seek each other out in order to hear and be heard. When one person tells another about some problem, both feel an emotional response. Listening becomes a sharing process, without criticism, and often without any specific solutions considered. At yet other times you listen for *information*, in getting directions or attending lectures. Listening for information requires more attention than does listening for entertainment and less emotional involvement than does listening to help a friend. Information is given so that you can translate it into action at some later point. Finally, there are times to listen for *analysis*. Here listening must be as acute as you can make it, for you are expected to take apart the messages you decode, criticize them, and respond to them. When you attend a movie to write a critical review of it for your campus or corporation paper, you need to expend a great deal of energy analyzing its different parts. Your companion could miss large portions of it while going to the refreshment stand, but you have to attend it all.

We do not imply that entertainment is the least important of the communicative purposes. Not at all. When you have worked hard and need a break, seeing a comedy at the movies might be the best thing you could do for your own well-being. Nor are we downplaying the need to share feelings with another. Both sender and receiver can derive great therapeutic satisfaction from simply having someone to listen when the other needs to talk.

5. Remember That the Process Uses Faulty Language Tools In Chapter 3 we discussed the problems inherent in language systems. Our discussion was not meant to discourage you from working to improve the use of ordered stimuli. People do communicate, often through symbol systems of great beauty and clarity. If you keep in mind the potentials for loss of meaning, the decoding problems can be controlled.

Languages, we have said, cannot accurately express an ever-changing world. If the people who use languages remember that discrepancy, communication can improve. In a verbal communication, a totally individual person decides to use the language about a changing world on a particular level of abstraction. From an almost limitless number of ways to express something, this abstractor selects one way. Since the meaning of the message is unique to the sender-abstractor, the most that can be hoped for is that the meaning of the message received is similar. Remember that *similar* cannot equal *identical*. Finally, be on guard for confusion between fact and inference. Do not allow language to trick you into insisting on more than you perceived.

6. Resist Assuming You Know What Will Be Said Before It Is Said Think of all the times you answered something before the other person even finished the original statement. That can be extremely irritating to your partner and it might be devastating for you to hear, "But that's not what I was going to say." Exercise patience. Even if you are sure you already know the ending, allow the sender to finish. One engineer learned this lesson to his embarrassment and almost at the cost of a contract. His entire department was preparing for a difficult customer progress review. He anticipated hesitation and resistance to certain design changes that had been made. Sure enough, the customer raised one of the objections. Interrupting the customer's explanation, the engineer went on to justify the change at length and expanded the discussion to explain every other change that had been made. The customer finally got a word in and said, "I was going to say that I see why that change had to be made. What I didn't realize until you told me was just how many other parts were altered. I think I'd better review this whole proposal again before the contract is signed." Fortunately, everyone was satisfied in the end, but the engineer learned an important lesson: do not anticipate what you think someone else will say.

7. Look for Entropy Sources If you accept the idea that entropy is always present in the communication process, you can prepare for it, identify it when it surfaces, and take steps to keep it at manageable levels. Message translations are never exact, so keep alert for the factors that garble them beyond any recognition.

Considerable courage is often required to recognize your own personal prejudices and mind-sets. We seldom like to admit that we really have problems listening to a particular kind of person—perhaps a woman with too much makeup; someone with a southern, British, or "foreign" accent; someone who speaks ungrammatically (or too precisely); a child; an old person. The list is endless. Rather than take a defensive stance regarding your own feelings about others, accept the responsibility of identifying such feelings and recognizing how they might interfere with your listening.

A bit of folk wisdom used to be posted in elementary school classes. It said, "What you are speaks so loudly that I cannot hear what you say." The words were intended, we assume, to inspire children to be worthy persons. That might be, but we are urging the same thing from the receiver's point of view. No matter who is speaking, the listener has the duty to identify personal entropy potentials and diminish them. It is never enough to use the alibi, "That's not my kind of person, so I didn't listen." If you are in the communication, all of the problems and their solutions apply to you.

8. Provide Honest and Clear Feedback We have discussed feedback as a monitoring system. Like a thermostat used to keep a room at a given temperature, feedback has to be accurate. If a thermostat signaled the furnace that the room was at the comfort range when, in fact, the room was freezing, a sensible home-owner would throw the mechanism out and get a more accurate one. As listeners, unfortunately, we often lie about our reactions. Audiences appear alert and interested, when they really think the speaker is boring. At parties, people smile and nod, when they actually disagree. We encourage communication behaviors that ought to be eliminated; and, therefore, the speaker is not able to recognize the problems that exist.

Granted most of us send some of that distorted feedback because we really do not care enough about the other person to help change behavior. It may also be a deeply felt sense of politeness. Who wants to hurt another's feelings? Perhaps fear is the cause. Who wants to tell the professor (chairperson of the board) that the remarks make no sense at all? Whatever the reason, the sender is getting positive reinforcement and will probably continue communicating in the same way.

Perhaps you should make a bargain with yourself: be aware that feedback is a continuous process. When you are in communication with another and the results really matter, send honest feedback clearly enough for the other person to understand most of what your reaction has been. How sad it is to hear of relationships ending with the suspicion that each person never knew what the other thought and felt.

9. Know and Prepare for Your Personal Entropies Because people differ, the attributes that keep one person from effective listening might be no problem at all to another. Consider the following list. Within it, there is probably something that gets in the way when you decode messages. It could be

- Things you do not like to hear
- Things with which you do not agree
- Things you see as difficult
- Things you see as boring
- Things you think you already know

Most of us have problems listening when we prejudge an event. Perhaps the material frightens us by its complexity, or angers us by its bias, or contains conclusions we wish were different. Whatever the reason, we have a harder time attending to what a sender says and does than when we find the material novel and exciting. One of us remembers a splendid teacher whose specialty was classical Europe, yet she attended most meetings of the science lecture series. When she was asked why she went to lectures so obviously out of the range of her interest, she said, "It teaches me discipline." Most of us could do with some of that discipline when we are faced with listening situations that we know will involve heavy personal sources of entropy.

Two Checks for Good Listening

By now you know that there are no easy steps in the communication process. Even when you work at communicating, losses always occur. Constant care is necessary and sometimes the benefits are hard to see. Each of the preceding nine suggestions will help you improve your listening, but the effort is long and strenuous. Two techniques seem to bring easily recognized results. They are not cure-alls for listening faults, but they do identify quickly where some of the faults might be. Try them; they will encourage you to keep working on the remaining nine.

Paraphrase is a simple, yet extremely effective device to tell the speaker (sender) what you thought was being said. You take anything the person sent and restate it in other words. If you remember to include the nonverbals, you can give the sender very accurate information. If you were a supervisor, how might you say "I will not tolerate your being late one more time" to your subordinate? It could be

- "I see your car didn't start again this morning. You are planning on getting a more reliable one, aren't you?"

  or
- "We had our regular 8:00 A.M. meeting without you. Again."

  or
- "I guess you're one of those night-people types. Too bad this is not a night business."

If the subordinate thought you were talking about cars or meetings or night people, your message would be lost. If, however, the listener paraphrases your statement, the response might be, "You are very irritated with my lateness, aren't you? You're telling me it has to stop. Right?" Try some paraphrasing. A good beginning is to ask, "What is this person really saying?"

A second fairly simple listening suggestion is *overt checking* of what the speaker has said so the speaker is satisfied that the message is accurate. This device is very effective in groups or whenever statements pass back and forth quickly. The rules are few. You may say anything you wish, but first restate what the previous speaker said to the speaker's satisfaction. In a workshop setting, successful adults find that they have to say and re-say something many times until the sender agrees that the message is now decoded accurately. If you watch most groups in action or most couples in argument, you are tempted to stop the whole process until participants discover whether they are all talking about the same thing. Think how much time would be saved if two technicians in an argument over some laboratory procedures would say, "You mean that the first step ought to have two peer check-backs, while the final step needs at least three?" Most people never know that they might be misunderstanding what the other is saying, rather than disagreeing with it. They do not attempt to check what they thought they heard.

WRAP-UP

The listening aspect of communication and its interplay with subsequent feedback completes the upper portion of the communication model. Listening failure is often an overlooked entropy source, yet its costs to personal and corporate misunderstanding are high.

Listening involves two levels: monitoring the self and monitoring others. In either form, feedback possibilities arise, and these make it likely that the successive messages might be altered.

There are reasons for poor listening, just as there are means of training oneself to become effective. We have suggested nine approaches to keep in mind and work on. In addition, two "quick techniques" of paraphrase and overt checking can be used. Learning good listening habits is a valuable professional and personal skill. Wendell Johnson comments on what each of us takes in and gives out in the listening process:

The kind of world a youngster gets inside his head may or may not be a good one in which to spend the rest of his life. At any rate, the kind it is, and is to be, depends on the questions the child asks and the answers his elders give him. It is largely their answers that he takes in and puts together to make his private world.[13]

For all people, this "private world" determines how they will interact with the world outside.

ACTIVITIES

1. During the next week, select a person you see daily and work to improve your listening with that person. Keep a log of how your attempts at improvement worked or did not work. Did the person note anything different about you?
2. To test the accuracy of your listening, work in the paraphrase device "Are you saying that . . . ?" into as many of your conversations as you can over the next two-week period. How many errors did you catch? How did people react to your attempt to understand them?
3. In class, convene a group of six persons and sit them facing each other in two straight lines of three people each. Tell them that they are to be on opposite sides of some controversial issue. One person from Side 1 speaks, advocating a reason for the position that side holds; then a person from Side 2 speaks in response to the first person, telling why Side 2 objects to the position; then a second person from Side 1 speaks, advocating another reason for the position; and so forth until each person has spoken, always alternating sides. Nobody may speak, however, before restating what the previous speaker has just said so that speaker is satisfied with the accuracy of the statement.

   What do the corrections and objections that will probably take place tell you about how much of what you heard is not what the speaker thought was said?
4. Identify your own listening entropies by keeping a log of your listening losses for a week. Note especially:

   - Where and with whom did the most serious losses happen?
   - How did you become aware that there were losses?
   - Did the losses form any kind of pattern?

5. Select any of the nine suggestions made in this chapter for the improvement of listening. Use these suggestions consciously to improve your losses as identified in Activity 4.

NOTES

1. Douglas Adams, *The Hitchhiker's Guide to the Galaxy* (New York: Harmony Books, 1979), 68.

2. Paul Tory Rankin, "The Measurement of the Ability to Understand Spoken Language" (Ph.D. diss. Univ. of Michigan, 1926), *Dissertation Abstracts* 12 (1952): 847–48.

3. Ralph G. Nichols and Leonard A. Stevens, *Are You Listening?* (New York: McGraw-Hill, 1957), 6–8.

4. Florence I. Wolff, Nadine C. Marsnik, William S. Tacey, and Ralph G. Nichols, *Perceptive Listening* (New York: Holt, Rinehart & Winston, 1983), 3.

5. Paul T. Rankin, "The Importance of Listening Ability," *English Journal*, College Edition, 17 (October 1928): 623–30.

6. Wolff, *Perceptive*, 29–30.

7. Raymond Ross, *Speech Communication Fundamentals and Practice*, 6th ed. (Englewood Cliffs, N.J.: Prentice-Hall, 1983), 92.

8. Ralph G. Nichols, "Listening Is a 10-Part Skill," *Nation's Business* 45 (July 1957): 56.

9. Wendell Johnson, *Your Most Enchanted Listener* (San Francisco: International Society for General Semantics, 1956).

10. Johnson, *Enchanted*, 23–24.

11. Larry Barker, *Listening Behavior* (Englewood Cliffs, N.J.: Prentice-Hall, 1971), 61–66.

12. Anita Taylor, Teresa Rosegrant, Arthur Meyer, and B. Thomas Samples, *Communicating* (Englewood Cliffs, N.J.: Prentice-Hall, 1977), 218.

13. Johnson, *Enchanted*, 9.

Summary: Part I

Fundamentals for Technical Communication

We began our discussions with two assumptions, both closely related:

- Most technical professionals function in corporate settings, some large, some small. These technical specialists and the people who work with them in national or international industries must understand each other's orientations to work together harmoniously.
- Technical professionals live in a communicative world, and their competence in communication helps or hinders their work and their dealings with coworkers.

To facilitate understanding of the communication process, we proposed a model that postulated a set of constants present in all communication. Inherent to the model is the concept of entropy which, as in the physical world, represents a direction toward a state of increasing disorder.

These communication constants are small in number and undergo transformation at each separate communication. As a result, no two communications are ever identical, although the same constants are present. Sources of the transform can be traced to differences in experience, the uniqueness of an individual perception, and professional expectations. The verbal definition of communication that we have proposed is *the sending of stimuli through some channel so that a receiver translates the stimuli into a message and then feeds back additional stimuli as a message monitor to the sender. In communication, as in the physical world, entropic noise is an inherent part of the process.*

The sender portion of the model involves the transmission of random and/or ordered stimuli. Random stimuli (signs) and ordered stimuli (symbols) serve as the raw material from which messages are selected and decoded.

Nonverbal stimuli—sometimes signs, sometimes symbols—are one kind of raw material from which messages are created. Different subcategories such as paralanguage, objects, color, touch, eye contact, space, and movement are used to code and decode messages.

Verbal stimuli usually appear as symbols. Using the work of the engineer/mathematician Alfred Korzybski and his General Semanticists, we discussed language as a system also subject to communication loss. The problems of relatively static language in a constantly dynamic world, differing levels of abstraction, and fact-inference confusion raise the probability of serious entropy losses. Verbal languages usually are transmitted along with one or more of the nonverbals, and so the probability for loss grows.

The function of decoder in the model belongs both to the sender and to the receiver. Called listening and/or perception, this portion of communication is the most heavily used skill. It is also performed with least efficiency and usually with no training. Differentiating listening from hearing, we described nine techniques for better listening and two methods for checking the accuracy of listening.

The model, its vocabulary, and the sender and receiver functions illustrate an extremely complex process. Simple solutions to communication problems and formula approaches to improvement do not exist. Understanding the system, however, allows a person to identify possibilities for entropy and keep them under control. For the technical student, such system analysis ought to be familiar. For the nontechnical person who will work with these professionals, understanding the communicative process promotes a common ground. For the technical professional in the corporate world, control of entropy is the key to competent career and personal communications.

PART II

PREPARATION FOR ENCOUNTER

CHAPTER FIVE

INTRAPERSONAL COMMUNICATION

Socrates, the central character of the Platonic dialogues, describes a critical incident that occurred in his youth. He journeyed to the temple of Apollo, where it was traditional to put a question to the Delphic Oracle. Socrates asked the Oracle where he might find the wise man, for he wished to study with the person of wisdom. The Oracle said that Socrates himself was the wise man. The young Socrates questioned the answer, doubting his own wisdom; but the Oracle gave him a charge that haunted and occupied him the rest of his life. He was ordered, "Know thyself."

Many centuries have passed since Socrates struggled with that answer, but the advice is still appropriate. Have you never wondered about the person who lives in your own skin? Most of us speculate why and how that person operates. We try to figure out the different pieces of the puzzle. From it all comes a picture of ourselves that the psychologists call *self-concept*. Much has been written about this picture; for, in many ways, it is the basis of all other relationships. Communication with the self is the precursor to all communication.

Knowing the self is not easy, and the picture you carry of yourself is very different from the one others hold. Perhaps you recall a conversation with a friend when you were stunned to realize that the friend saw you as very different from the way you saw yourself. Many audiences would marvel to know that speakers who appear so poised and sure of themselves while delivering a difficult presentation, have spent the previous day convinced that they were incapable of uttering two consecutive sentences.

The private messages you communicate when you are both sender and receiver are called *intrapersonal communication*. These messages are, literally, the endless stream of stimuli you decode with and within yourself. In Part I, we discussed the ever-present signals passing from organ to organ, the constant thought patterns that consciously or subconsciously occupy your mind, the processing of limitless sounds and images, and your attendance to yourself as your own listener. These activities constitute your intrapersonal communications.

Sometimes conversations with the self can be hurtful. People can, after all, talk themselves into damaging behaviors. You can tell yourself that you are hungry for a piece of chocolate cake, even though you have just finished a large meal. You can tell yourself that talking to a group of people is threatening, while talking to them one at a time is not. The fear of public speaking, usually called stage fright, can be extremely destructive, and we will try to defuse that fear through understanding. This chapter deals with the constituents of self-image: human values, human needs, and human anxiety. In Chapter 2 if you recall, we described *background* in the communication model as everything that has happened to both sender and receiver up to the moment of transmission. Unless we understand some of what went into the sender's and receiver's conversations with the self *while* events in their past occurred, a vital piece of the communication process will be missing.

THE VERSIONS OF ME

The *I* who forms the *me* of self-concept receives both internal and external input about the self. Our heredity gave us the raw materials for what we were to become. From the first day of our lives, however, each of us had different amounts of handling, talking to, listening to, words put into our heads, places we have been, and things we learned to do. If we were touched gently, by someone who wanted us there, we concluded that the world was a caring place. If few people spoke to us or spoke in irritation, we found an indifferent world. If people told us that boys were not supposed to cry or that girls were incapable of doing math, we ended up with words in our heads that some of us can never get out. If we went to the zoo or learned to kick a soccer ball, our world expanded. Throughout it all, the persons we are took in all that information, decoded it, and formed our self-images as citizens of that world.

As with every process, our image of the self evolves and changes rapidly. In both college and corporate settings, we make numerous alterations, probably daily. Think of it as a series of massive communication systems: dozens of persons doing and saying things to us. We take in the stimuli and decode them into messages that, in turn, are matched to past messages we have about ourselves. There are few more humbling realizations than those made in the early days of college or a new job. A typical freshman reaction to an F on an assignment might be, "I've never had an F on a project before, and I worked harder on this than I ever did on anything in high school. I guess I'm not such a hotshot." The newly hired professional suggests an idea during a staff meeting and is dismissed with a patronizing smile. "I thought it was a brilliant idea. My microbiology teacher said it was when I mentioned it in class. I guess these people have heard it all a thousand times." From the pain of such experiences that, we hope, is balanced by other messages of bright promise and fine potential, we refine the picture of ourselves.

How can there be conflicting versions of the same person? Isn't there a single *real* me? Such questions have no easy answer, of course. Psychoanalysis and much self-help psychology attempt to restructure a self-image so that the public picture improves as well. Relying again on your training, which is one of your background components, we offer the model shown in Figure 5-1 to help you account for the different interpretations of you.

The model suggests that any person is seen in at least five different versions. Each interlocks with the others, affecting and being affected by them. Each is also in process, changing constantly. Central to this theory is the realization that our interaction with others today draws from past interactions (our backgrounds). These, in turn, become part of our background and feed forward into communications in the future. In a very real sense, each of us is continually coming into being as a result of the communications we have experienced.

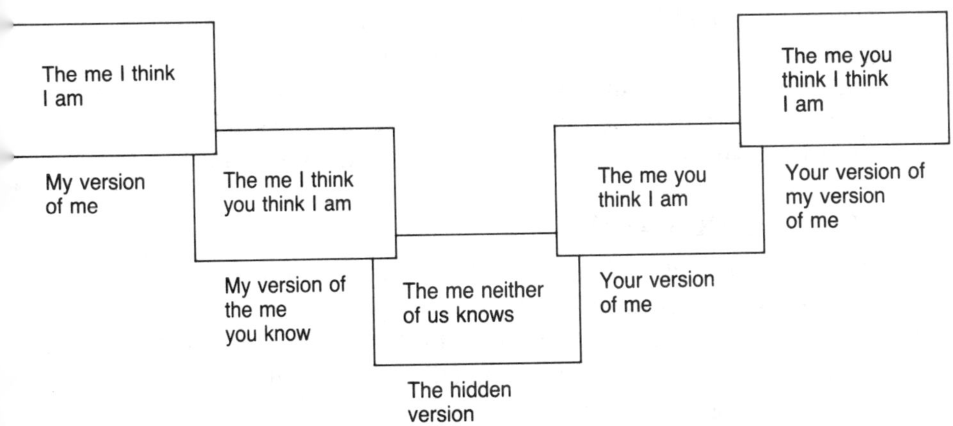

Figure 5-1 The Versions of Me model.

My Version of Me

The first box in Figure 5-1 represents the self from the person's own point of view. It is the you that you yourself define. This version of personhood has its genesis in childhood. It is redefined as significant others tell you about their versions of you.

My Version of the Me You Know

If you made a list of what you thought every other person in any group thought you were like, what would that list contain? Adjectives such as *friendly* or *quiet?* Nouns such as *clown* or *loner?* Would your list change from group to group or remain consistent? How would it change if you made the list for your family or close friends? The experience might be a revealing one, for you would be dealing with the inferences others hold about you. Your list might be accurate, but it would contain many errors too. You think your interviewers saw you as "sophisticated"; they actually perceived you as "phoney." You think your department coworkers regard you as "thorough." They feel you "put things off." *Same* person? Hardly.

The Hidden Version

You have many capabilities that neither you nor your observers realize. We have chosen to call these the hidden version of the self. Included are qualities you simply never recognized in yourself. Stories surface about heroic actions, and the hero says, "I never knew I could do something like that." Terrible crimes are committed, and the accused says, "I never knew I could do something like that." Often, the communities these people live in agree: "Who would have thought that laid-back old Gerry would jump into twenty feet of water to rescue someone?" "Who would have thought that quiet Lou could stab someone to death?" Who, indeed? Possibly Gerry and Lou were just as surprised as their neighbors, for many things about any person are unknown, yet are part of that person just the same.

Your Version of Me

Just as you could mistake another's picture of you, so that other person can be mistaken about you too. Students say that their parents refuse to see them "as they are." Perhaps a mother and father are used to thinking about a child's needing protection, while the "child" is a strapping adult capable of independence. Sometimes the version of you that people adopt has little to do with what you yourself are like. People have the dangerous habit of assuming that a whole group is identical, so any member of that group can be stereotyped. Quickly, now: imagine a nuclear physicist. Is it a he or a she? Does the person wear glasses, talk in technical jargon, stay away from cocktail parties, like to play computer games? Because you cannot visualize people in all the different shapes and personalities that personify a nuclear physicist, you have stereotyped. To the degree that your receivers think in stereotypes about the technical discipline you plan to enter, this mistaken version of you will affect your respective communication processes.

Your Version of My Version of Me

Not only do people stereotype, but they even assume that they know what the members of a group think about themselves. One of us recalls hearing two young women talk about a fraternity. "Don't bother going to their party. Most of the brothers are computer majors, and all they want to talk about is machines. They think they're so smart. You'll be bored out of your skull." It would probably have been pointless to tell the speaker that she could not possibly know what another's self-concept is, let alone know the self-concepts for the many individuals in a fraternity. To her, all communications with a computer major would include her version of another's self-concept.

Thinking about the self in the framework of these possible versions allows you to recognize many potential problems. As long as the different versions are in harmony with each other, no difficulty is apparent. If (1) you see yourself as outgoing and eager for life, (2) that version matches what you think others say about you, (3) they, in turn, assess you that way, (4) they also claim that is how you see yourself, and (5) that seems to be the way even your hidden capabilities shape up, your internal and external feedback will be consistent.

Suppose, however, that the different versions vary significantly. You think you are outgoing and eager for life, but you suspect others might think you somewhat cautious. They, in fact, think of you as loud and picky, while they figure your picture of yourself is that of a know-it-all leader. Neither you nor they know that you are secretly frightened of being rejected and so use abrasive humor to hide your fear. If this *were* you, you would have a considerable problem. Your messages to yourself are skewed, and your signals to others are just as entropic. You could dismiss the confusion as a "communication problem," which it is; but, until you understand that messages are being misinterpreted in every encounter you have, you are not likely to work on the entropy.

Neither does it seem to help much to wonder about which is the *real you.* Each version is real. Each is the result of your own intrapersonal communication or the result of communications sent by you and decoded by another. Rather than wondering whose fault it is and how the *real you* can surface, pay attention to the systems approach of communication. Messages are going astray. Look for the sources of the entropy and work on them. Accepting feedback, without being defensive about it, can help you learn a great deal about those versions of you others have. When the versions are all fairly consistent, but never *identical,* you have attained a match in your self-image and public picture.

HUMAN VALUES AND NEEDS

The complexity of intrapersonal messages makes a fascinating study. Consider for a moment how you talk to yourself when you make decisions or weigh alternatives. *While* you are telling your roommate that you cannot go on the ski trip this weekend because of the test announced for Monday, you are carrying on a steady conversation with yourself. As your roommate gives you counterarguments, you consider them privately in your own mind before you voice an answer. Have you ever wondered why you come up with the decisions that you do? Some might claim that the roommate's reasons were so good that "I just couldn't say no." That evaluation is not quite accurate. You *judged* the reasons according to your own value system and, as a result of this inner communication, you acted. Two theories of what might be happening inside you provide insight to your behavior. These theories postulate an internal screen through which each communication passes. We are speaking of value and need systems, and they are part of the background of each human communicator.

Values

Each of us acts according to an unwritten list of judgmental choices we would be likely to make in a given situation. The list is an ordered one; some things would always be chosen in preference to other things. People spend their lives refining the order. In many ways, the list differs from person to person, and the same person's choices might even change at different times of life. We are speaking about *values* when we speak of making choices according to priority.

Thomas Scheidel points out that some theorists try to categorize all persons into "Value Types."[1] Drawing on the work of the German scholar, Eduard Spranger, Scheidel claims that it is useful to isolate the predominant value system operating in any person. One individual might be the "Theoretical Type," responding to pursuits of discovery; another might be the "Aesthetic Type," choosing beauty and harmony over other possibilities; still another might be a "Political Type," valuing competition and personal power over other choices. These values, Scheidel concludes, partially determine what will influence a certain person.

Whether or not it is possible to categorize complex individuals into broad categories of type, examining your own value systems is an informative experience. Richard Chewning, writing about ethics and the business environment, warns that, "What we need to be consciously aware of is that all of our thinking and behavior are value-laden."[2]

From our viewpoint, these important personal and business choices are made in the privacy of intrapersonal communication. They, in turn, become a part of our background and come to bear on future communication events. Our communication model (Figure 2-11) suggests that background draws from an indeterminate number of variables. Each of these, values among them, determines the sort of stimuli, channels, and symbols we select in the transmission of every communication we send.

How do you find what your personal list of values might be? Try actually making a literal list. You could write down all the broad categories of things you think are important:

- Personal Peace
- Good job
- Success
- Many friends
- Etc.

You could also make the list very specific:

- Vice President by the time I'm thirty-five.
- A conscience free of guilt.
- An income of at least $500,000/year.
- Parent of two healthy children.
- Etc.

When the list is complete, ask yourself which item you would give up in order to get which other one. Eventually you will realize that one is more important to you than anything else. That is your top value, your highest priority. The remainder can be arranged in forced-choice order until you have a unique list that is right for you.

Closely related to values are ethics. *Ethics* are choices made through evaluation of good-bad morality questions. You might decide that the value of an income of one-half million dollars can only be reached by some dishonest action or at the price of harming your competition. Your ethics dictate that neither of these actions is morally good, so you decide you cannot work for that value. For most of you in college, education is a costly value. For some, staying in college involves serious ethical questions. Do I cheat on the test in order to maintain the required grade-point average? Do I get an abortion in order to remain in school? Each of us considers these questions in the light of the value systems we carry and the ethical standards that help us decide. Depending on how we talk to ourselves in the private conversations of intrapersonal communication, we make choices.

Needs

For the last thirty or forty years, psychologists have postulated sets of needs that act as driving forces in human actions. Perhaps the most famous needs theory is that of Abraham Maslow.[3] His system, usually represented as a triangle, has been reproduced widely; it permits broad interpretation, yet it is simple to understand. See Figure 5-2. Basically, a *need* can be defined as a state of unbalance in any system: in its mild form, need will frustrate the system; and, in its most severe form, need will shut down the system. The most common interpretations of Maslow suggest a five-level set of needs, each less pervasive than the one below it. Unless the lower need is generally satisfied, the higher one does not come into play. When the first two are satisfied, the third one becomes pertinent, and so forth. In this way, the Maslow needs form a hierarchy.

Maslow's first-level need, the physiological or body needs, are usually thought to be food, water, air, sleep, warmth, and sex. Each of these permits the species to continue. Withholding any single one of them from the entire species means extinction of that species. If there were no procreation, for example, it might take a hundred years or so for the last of us to die and humankind to disappear. If there were no air to breathe, it might only take four or five minutes, but the

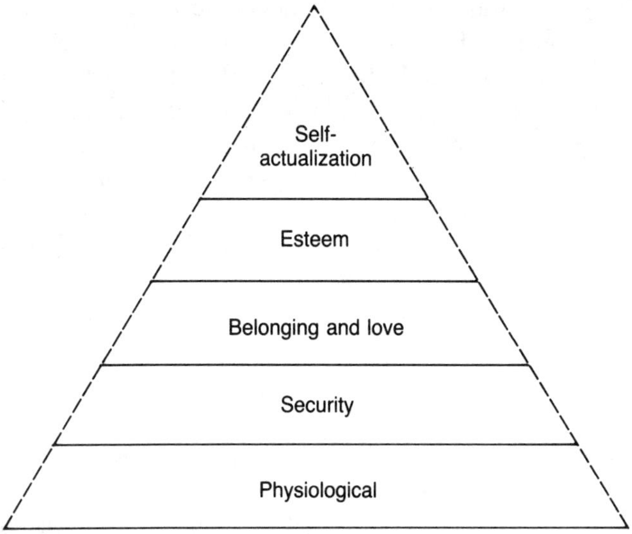

Figure 5-2 The Hierarchy of Needs model, based on work developed by Abraham Maslow.

result would be the same. As a consequence of their importance to survival, these needs are very strong. The desperately thirsty person becomes obsessed with finding water and does not worry about being accepted or loved. The moisture requirements of the body are in a state of critical unbalance. If imbalance continues, the system shuts down and the person dies of thirst.

As a human advances to the higher needs, direct shut-down might not occur, but the will to keep living ceases and sometimes suicide kills the system anyway. The second-level safety needs are those that keep a person relatively threat-free. Desires for predictability and order are part of this need. Imagine a world where everything was utterly random. Nothing could be planned; nothing could be forecast. One moment your room is there; the next it is gone. The frustrations are easy to predict. In a state of extreme randomness, people would become so disoriented that they would probably not want to survive.

The third level is the need to be accepted, to belong somewhere, to find someone to love. Our species is a social one, and when its members think that there is no place or person they can belong to, the need state becomes severe. Try reading some of the advice columns in newspapers. Some of those letters are cries of pain. The beloved is gone, and another will never be found. The writers say they do not care to live any longer. For most people, of course, another loved one is found, and the need state returns to balance.

The fourth level is the need for esteem, respect for oneself, and the respect given by others. We want to win and have our observers cheer. We want to have something we are proud of. People buy fancy cars that are hard to park or clothing they cannot afford. They fantasize about showing up at a posh party with the most attractive date in the room. They train their bodies to the extreme so they can be cited in the record books for shaving one one-hundredth of a second off a standing record. The agony reflected on the face of an Olympic competitor who did *not* set that new record is moving proof of the depth of this need.

The fifth need, in many ways, is the most elusive. Maslow calls it self-actualization. It is the self-defined feeling we have when we have achieved all that we set out to do. We reach self-actualization when we have completed all the goals we set and are satisfied that we are all that we can be. Needless to say, not too many of us are such people. Maslow actually postulates two needs beyond self-actualization, but these take us into controversial areas. Our discussion stops at self-actualization.

Needs, like values, are recognized and dealt with in the conversations people have with themselves. Some psychologists claim that the importance of needs is that they result in actional behaviors.[4] For example, the hungry person looks for food; someone seeking love will go where a love-object might be found; the esteem-seeker will put on the white lab coat to be recognized as one of the prized technical staff.

The conversations you have with yourself while you are in one or another of these states of need certainly involve your values and perhaps involve your ethics. If you are that hungry person, you might choose not to steal the food because you value your unblemished record more than relieving the need. You might also decide not to steal because you find theft morally objectionable. But acute hunger becomes a serious and even tragic problem. Stories of the cannibalism that occurred when a plane crashed in a deserted wasteland horrify us, as well they should. The conflict between the extreme need that motivated such action and the value system that could not prevent it tells us a little about the private conversations of those who had to make such choices. In an affluent and, we hope, disaster-free life, intrapersonal communications are usually less wrenching.

DEALING WITH ANXIETY

The importance of intrapersonal messages reaches beyond "know thyself," to all communication settings. We now arrive at a topic certain to come up during any class studying oral discourse. It is called by different names: stage fright, speech anxiety, performance apprehension, reticence, and—most recently—quietness or shyness. It refers back to that stereotype of a platform speaker with shaky hands, ragged breath, red face, and blank mind. People who study speech hope that a fringe benefit of their study will be control over those unpleasant symptoms.

Speech Anxiety

Most of us know that speech anxiety, the name we prefer to use, is widespread. In 1974, *The Bruskin Report* startled and amused many people when it concluded that "speaking before a group" topped the list of adult fears.[5] It preceded financial worries, loneliness, and even death. Linkugel and Berg note that "The basic human tendency, and it's pretty much universal, is to talk."[6] Yet, despite this "normal" pattern, people report terrible fear when they have to face a group.

Perhaps you have watched the television broadcast of the Motion Picture Academy Awards and were surprised at the nervousness displayed by some of the big-name stars. You might understand it in those nominated for the awards. They are, after all, under tremendous strain. How, then, do you account for the bumbling words and trembly voices of the presenters? Here are the industry's image-setters—the perfectly macho man, under control even when he is pinned down by gunfire of superior forces, or the most admired woman, cool even as she walks knowingly to certain death—and their hands tremble so badly that they cannot open the presentational envelope. Obviously, even the great suffer from the pangs you thought talent and training eliminated.

Imagine yourself, then, about to answer questions about your technical report before the general manager's staff or waiting your turn to present that classroom speech. You are scared, and how can you cope with your fear? Some people refuse to discuss it on the assumption that if you ignore it, it will vanish. Some people tell themselves that they feel nothing, as if lying makes it better. Some give in to the feeling, avoiding any situation that might cause it. None of these techniques works. The feelings are there, and neither ignoring nor denying them can alter that. The educated professional cannot avoid oral communication, so sometimes you are going to have speech anxiety. Our approach is to offer you some ways of understanding what is happening. As you acknowledge your feelings and learn to use them, the fear is defused and the situation returns once more to your control.

Fear

Many of the terms used to describe speech anxiety are just as ambiguous to psychologists as they are to communication theorists studying this behavior. Some distinguish between *fear*, as a specific reaction to a specific event, and *anxiety*, a generalized reaction to the unknown. Fear, then, would be your reaction to taking Professor Zalowiek's midsemester test in Organic Chemistry II, and anxiety would be your reaction to taking tests in general. Others use the two terms interchangeably, along with worry, apprehension, dread, and misgiving.[7] Surfacing in the speech-communication literature are studies that compound the problem further by linking speech anxiety to shyness and quietness.[8] We shall try to sort out the complexity by first taking a look at fear, and then seeing where linking shyness or quietness to speech anxiety might lead.

Fear is an emotional state. We leave it to the psychologists to decide if it is identical with anxiety or a special case. For our purposes, they are the same. Especially in speech anxiety, the symptoms of the anxious and the frightened are measured in the same physiological reactions.[9] There is increased heart rate and breathing, sweating, shakiness, and stomach distress. Some threat presents itself to the organism, and it reacts by displaying these symptoms. Then what? How can a creature, specifically a human, cope with that fear?

The creature might:

- Flee.
- Stand and fight.
- Surrender to it.
- Escalate it.
- Ignore it.
- Redefine it.
- Recognize and use it.

The first three options are open to all creatures; the last four are probably possible only to humans. A deer or human could run from the threat, or it could turn and attack. Wolves are said to turn belly-up when the fight fails, signaling acceptance of surrender. They are not any less fearful of the opponent who has just beaten them, but they accept the fear and its consequences. People also accept their fears and learn to live with them. Tragic stories of abused children teach us, among other human horrors, that our species learns to adapt to the most trying circumstances. When there is no way out of the pain and fear, these abused children learn to live with it.

The last four means of dealing with fear seem peculiarly human, probably because each of them requires sophisticated language use to make it work. People can talk themselves into escalated states of fear. Jack R. Gibb says that, "Fear and distrust overperceive the danger, trigger defensive behavior in others, escalate the tension, and are self-fulfilling—that is, *fear creates the danger*."[10] Speakers seem to fit Gibb's description well. They worry about past experiences with symptoms; they foretell awful things happening when they attempt to speak during the group discussion; they feed the fear with intrapersonal horror stories. They often succeed in taking a manageable anxiety and escalating it out of control.

People are also capable of lying to themselves quite effectively. They ignore what their bodies are telling them, repress recognition of what they feel, and cope with fear by pretending it does not exist. One of us remembers a tall strapping male student who told a class that he never experienced any speech anxiety. He said he could not understand how others let themselves get into such states. At his next presentation, his face was extremely pale, his body was stiff with stress, and his words were distorted through a tightly clenched jaw. At a subsequent meeting with the teacher, he again denied feeling any fear. When she told him that the person she was used to seeing in class was animated and anything but stiff, he could not accept her comments. "I'm exactly the same as I was during the speech. I felt relaxed. I liked what I was doing." The young man's communication with himself was obviously distorted.

One magic quality of language, discussed in Chapter 3, is its ability to redefine anything as something less fearful. A colleague once told us that she deals with her speech fears by having a long talk with herself. When she knows she has prepared, she said, then the fears have to be the result of "irrational expectations." What she meant was that when you have done all that is reasonable for a sound preparation, what remains is to speak honestly with the person who lives in your skin. Ask yourself why you have defined this occasion as threatening. How is this time different from all the others? What will it matter, really, if the audience giggles when you forget the next sentence? Do you truly want your audience to hear what you have to say? If not, why not? If so, why are you so worried about yourself? This intrapersonal dialogue can set the entire encounter into another framework, one that you find acceptable.

Finally, a human can recognize personal fears, their symptoms, and how they might be used. Harry Weinberg says:

Do everything to change that which can be changed and accept that which at the moment cannot. The act of acceptance, the will to bear discomfort, has an immediate and profound effect upon the intensity of the pain. It becomes much more bearable, less threatening.[11]

To a large degree, speech anxiety is an intrapersonal miscommunication. For whatever reasons, those who experience it (most of us, in fact) are sending erroneous messages to themselves. Nobody is actually threatening us, and nobody will attack. It is not possible to stand and fight, and neither is it appropriate to flee. What remains is the use of language to activate a way of coping with the problem.

As you recognize what your own symptoms might be (and, by the way, they differ from person to person), you have a head start on coming up with solutions. Your body is preparing you for a high-energy-using flight-or-fight situation. You need to burn up some of that energy, or the extremities where running and fighting take place will be shaky. You need to regulate your breathing, consciously pacing it to the speaking rate rather than the running rate. You need to know that your body will slow down "unneeded systems" such as digestion in order to divert energy to those organs needed in physical trial. The heavy meal sitting in your stomach will remain there, giving you those feelings of stomach distress.

With your body under control physically, you can put all the extra adrenalin to use keeping the mental faculties alert and on target. Of course, the fear you experience might be a very accurate assessment of your readiness for the upcoming task. If you have not done adequate research, not planned a presentation to do justice to the subject and its audience, and not practiced, you deserve the feelings of inadequacy you have. You are simply not ready, and you should be anxious. As a responsible communicator, you must work to feel that you are:

- Physically ready with all extra energy expended
- Responsibly prepared with information and practice
- In possession of something worthwhile to say
- Engaging in a communication with one person at a time
- In the presence of supportive and understanding people
- Capable of talking to yourself accurately and honestly
- About to have a successful experience

Quietness and Shyness

Some people will do everything we have suggested in the previous section, and they will still be unable to feel comfortable in communicating. These persons sometimes classify themselves as "shy." Researchers have suggested that speech anxiety and shyness are "conceptual twins."[12] A debate has raged in the journals over whether they are the same or not, but even those who deny their identity admit that speech anxiety, shyness, and resultant quiet behavior are related.[13]

Several points are worth noting: People who define themselves as shy seem to be fearful of *all* communication situations, not just those involving public speaking. These people are called Communication Apprehensives, and they are fearful of and anxious about communication with other people in all settings. Furthermore, these people tend to be quiet, and the pattern is established in childhood.[14]

Several very serious results occur when shy people reach school age. They tend to be quiet, purposely avoiding verbal contacts. Their apprehension, however, does not solidify into adult norms until around the fourth or fifth grade.[15] James McCroskey found that some disturbing conclusions can be drawn about shy persons by the time they reach college: (1) They arrange their entire environment around the avoidance of interaction. They drop courses where contacts are expected; they avoid seeing advisors; they seldom seek help; they seldom say anything in class. (2) Others often react negatively to quiet persons, judging them as less socially attractive, less competent, less composed, less intelligent and less likely to be entrusted with leadership.[16] Consider what this does to the student participating in highly competitive professional disciplines.

The human costs to a shy and communication-apprehensive student moving through college are extremely high. They may be even higher in the professional world after college. McCroskey points out that quiet people are discriminated against in hiring; they tend to have lower paying jobs and lower status positions; they are less likely to be retained in their jobs when cuts must be made; and they are unlikely to be involved with political and decision making processes. He concludes: "people who are quiet also tend to be disadvantaged in many of the social, economic, and political aspects of their lives."[17] Ronald Adler's findings agree with McCroskey's, but he adds that "reticent" people are less likely to receive job interviews in the first place and few job offers. When hired, they prefer occupations that involve low interaction, and these freeze them into low-paying positions with little likelihood of advancement. Adler concludes:

Consequences such as these leave most apprehensive communicators with diminished self-esteem. They perceive themselves as relatively inadequate and worthless in social interactions, unaffectionate and unsympathetic, physically unattractive, overly deferent and conformist, unparticipative, friendless, and without strong social attachments. While these self-evaluations may be excessively harsh in many cases, the subjective experience of being reticent makes the phenomenon a serious problem.[18]

We have come full cycle back to self-concept. The unfortunate people in Adler's quotation might be none of the things they evaluate themselves to be. That hardly matters. Their unwillingness to deal with the miscommunicated messages they carry in their heads translates itself to avoidance behavior. Psychologist Lynn Alden of the University of British Columbia says that only about 5 percent of the population is habitually and cripplingly shy; but an estimated 40 percent describe themselves as such.[19] Further, Warren Jones, professor of psychology at the University of Tulsa, claims that if some of the words used on shyness tests are changed, the numbers of people reporting themselves as shy changes.[20] For example, when questions such as "Are you shy when . . ." are replaced with "Are you tense or anxious when . . . ," differences in shyness according to gender begin to disappear. Jones concludes that, "There are only slightly more shy women than men, and at the college level, no difference at all."

If Alden and Jones are correct, then the language we carry in our heads about our own Version of Me becomes extremely important. People convinced that they are "shy" can translate their intrapersonal message into actions of quietness and noncontact. Their receivers seemingly translate those behaviors into a string of negative impressions. The impressions fed back to the sender solidify the Version of Me. All the versions of the self become consistent, and both the public and private self becomes stunted. Only in the Hidden Version are there capabilities for something different.

If you are one of the "shy" ones, how might you deal with your fear? Again, there are no "Become Outgoing in Four Easy Lessons" answers. But you can improve your intrapersonal communication, and that new set of messages might make it easier for you to make contact with others.

Change the Labels You Use to Talk About Yourself People who call themselves shy *act* shy. "Of course I didn't talk to her. She's special, and I'm too shy to approach her." "I just don't make friends easily. Never could. I've always been shy." Instead of the *I am shy* language, move to a lower level of verbal abstraction such as:

- "I choose not to go into that room."
- "I believe she will not talk to me, so I elect not to talk to her first."
- "I accept (or do not accept) my feelings of shakiness when I have to go to Professor Washington's office."
- "I select not to place myself in the position of speaking in class."

When you are accurate in describing what you feel and what is happening, you begin to be more honest with yourself. *I am shy* language makes you a passive and helpless victim. "I *choose* not to do this" gives control of decision making back to you.

Stop Lying to Yourself About Your Feelings Who should know better than you what goes on in your own skin? Be honest with yourself. Do not deny what you feel. Accept it. If your stomach feels queasy before you walk into a room of strangers, recognize the feeling. Dealing with it means making the decision (*your* decision) whether you will bear the pain or run from it. If you choose to run, accept that too as a decision you made. Do not, however, blame it on "shyness" that is beyond your control.

Do Those Things That Frighten You into Shy Behavior Doing the impossible changes it into the possible. Estelle Zannes says that we often live up to our own image of ourselves:

If we perceive ourselves as being "quiet," we may not speak up at meetings. But self-concept can be influenced by others around us. For example, if we finally do stand up and speak, and our audience applauds us warmly, we may change our self-concept in that area.[21]

Doing something that frightens you and succeeding at it is a powerful experience. What makes it especially satisfying is the realization that *you* chose to do it despite feelings of discomfort. That takes courage and very careful communication with yourself.

Recognize Your Fear of Rejection Most persons dread another's rejection. Villard and Whipple say that this dread is perhaps the most powerful force in all interpersonal dynamics.[22] Shy people may decide that this fear is simply too devastating and cope with it by withdrawing from others. Villard and Whipple compare them to cowboys in old westerns who keep ducking behind boulders, putting their hats on sticks, and cautiously raising them to see if someone will shoot:

For we all maintain emotional bank accounts where the results of our interpersonal transactions are totaled and recorded. Those who have invested and "lost" find that they are fearful of investing again, yet they are the very ones who need the support and confirmation that such an investment might provide.[23]

Perhaps that is the key: be willing to risk again. Recognize your own fear of rejection. Know that the *identical* thing can never happen again, although a *similar* one might. Go for it, despite the fear.

One of us remembers an extremely capable job applicant for a senior engineering position. His résumé reflected a person of superb technical competency, so it was curious as to why he had not progressed much during his years of employment with a large firm. The interview provided the answer. The man projected a personality of near-total avoidance. His face seemed fixed in a permanent scowl; he would not maintain even minimal eye contact; and he answered all questions in monosyllables. The interviewing team tried all the techniques they knew to help him relax and open up. Nothing worked. Despite his being the most competent applicant they interviewed, he was *not even considered* for the position. At some time in that unfortunate man's background, he decided that anything was better than the threat of reaching out to another. The price he is paying for that erroneous intrapersonal message is an unfulfilled career. "You cannot not communicate" begins with the self, but its consequences color an entire life.

WRAP-UP

We maintain that knowing yourself is part of the intrapersonal message system. This system, in turn, structures all other communications. The concept of the self is extremely important in understanding other human contacts.

We postulate a Versions of Me model that points to the five different "real selves" that make up self-concept. The more similar each of these versions is to each of the others, the more accurate are the intrapersonal messages.

Understanding how the self is defined involves looking at the basis of values, needs, and fears. Speech anxiety, a special problem of fear through miscommunicated intrapersonal messages, can be handled through understanding the process. Communication anxiety, on the other hand, describes quite another system, one where communication is avoided in all situations. Such avoidance is associated with "shyness," "quietness," "reticence," and a string of negative evaluations by listeners. Research indicates that people who cannot cope with their communication fears face serious personal and career ramifications.

ACTIVITIES

1. Pick a half hour each day when you can be alone and are not likely to be interrupted. During that half hour, consciously engage yourself in internal dialogue and attempt to reach the goal set in any *one* of the following problems:
   a. Clarify something that happened during the day, so you feel you understand the event, the people who participated in it, and your actions while it was going on.
   b. Analyze your feelings about someone with whom you communicated that day. Try to understand why you have the reaction you do.
   c. Isolate some news event that happened that day. Decide what you know about it, what you still need to find out about it, and how it was that you learned about the event.

   Do not stop your consideration of *one* of the above before the half-hour ends.

2. Examine some recent full-page magazine advertisement. Identify the primary (most focused) need and any other need(s) the advertisers wanted to play upon. How did they work to accomplish their goal?
3. Make a list of your own values. Rank them so that you would be willing to give up the lower-ranked ones in favor of the higher-ranked ones. Which one are you not willing to give up for any other one? Does this or any of the others cause you an ethical concern?
4. Make a list of your fears. If speech anxiety is on your list, how serious is it in comparison with your other fears? If speech anxiety is not on your list, how do you account for your lack of fear during speaking?
5. If you classify yourself as shy, work with one or two of the suggestions made in this chapter on how to deal with shyness. After a fairly long period (a full semester, for example), has your perception of your own shyness and the way you have dealt with it changed your behavior? You might want to keep a log of your "feelings of shyness" episodes and how you went about dealing with them.

NOTES

1. Thomas Scheidel, *Speech Communication and Human Interaction* (Glenview, Ill.: Scott, Foresman, 1972), 219–20.
2. Richard C. Chewning, *Business Ethics in a Changing Culture* (Reston, Va.: Reston Publishing Co., 1984), 8.
3. Abraham H. Maslow, *Motivation and Personality*, 2nd ed. (New York: Harper & Row, 1970), 35–72.
4. John P. Houston, Helen Bee, and David C. Rimm, *Invitation to Psychology*, 2nd ed. (New York: Academic Press, 1983), 312–13.
5. "What are Americans Afraid Of?" *The Bruskin Report* (New Brunswick, N.J.: R. H. Bruskin Associates, July 1973), 1.
6. Wil A. Linkugel and David M. Berg, *A Time to Speak* (Belmont, Calif.: Wadsworth, 1970), 6–8.
7. Houston, Bee, and Rimm, *Invitation*, 356.
8. See James C. McCroskey and Virginia P. Richmond, "Communication Apprehension and Shyness: Conceptual and Operational Distinctions," *Central States Speech Journal* 33 (Fall 1982): 458–68.
9. Houston, Bee, and Rimm, *Invitation*, 356.

10. R. Gibb, *Trust: A New View of Personal and Organizational Development* (Los Angeles: The Guild of Tutors Press, 1978), 16.

11. Harry Weinberg, *Levels of Knowing and Existence* (New York: Harper & Row, 1959), 188.

12. Malcolm R. Parks, "A Test of the Cross-Situational Consistency of Communication Apprehension," *Communication Monographs* 47 (1980): 220–32.

13. McCroskey and Richmond, "Communication Apprehension," 460.

14. James C. McCroskey, "Quiet Children and the Classroom Teacher," ERIC Clearinghouse, Speech Communication Association (November 1977), 7.

15. James C. McCroskey, "Quiet Children in the Classroom: On Helping Not Hurting," *Communication Education* 29 (July 1980): 239–44.

16. McCroskey, "Quiet Children and the Classroom Teacher," 10–11.

17. McCroskey, "Quiet Children and the Classroom Teacher," 11.

18. Ronald B. Adler, "Integrating Reticence Management Into the Basic Communication Curriculum," *Communication Education* 29 (July 1980): 213–31.

19. Lynn Alden, quoted by Lindsy Van Gelder, "Social Anxieties," *Ms.* (May 1985), 118–19, 148–49.

20. Warren Jones, quoted by Lindsy Van Gelder, "Social Anxieties," *Ms.* (May 1985), 118–19, 148–49.

21. Estelle Zannes, *Communication: The Widening Circle* (Reading, Mass.: Addison-Wesley, 1982), 20–21.

22. Kenneth L. Villard and Leland J. Whipple, *Beginnings in Relational Communication* (New York: John Wiley and Sons, 1976), 179.

23. Villard and Whipple, *Beginnings*, 181.

CHAPTER SIX

FROM SELF TO OTHERS: AUDIENCES

Communication does not take place with an *audience*. The word calls up connotations that obscure all we have been saying about the process of sending and translating stimuli. The danger lies in the assumption that audiences function as a unified whole: it (the audience) responds; it (the audience) acts. The concept is inaccurate. Communication always happens with a *single* receiver. The receiver decodes the stimuli sent from the viewpoint of a unique, personal background. A hundred or a hundred thousand such receivers might be in the same place listening to the same speaker, but every one of them completes a process of reception different from every other one. An *individual* might be affected by the experience of sitting in the same place with many other people. That person might be affected by those other people and might decode the message differently if alone or with one other person. Nevertheless, in the final analysis, it is a single person's reaction that completes the transmission process. We will use the word *audience* throughout this book, but remember that we mean by it an individual receiver.

Every receiver you encounter is audience. The friend with whom you have coffee every Thursday morning, the interviewer or panel of interviewers who decides on your job offer, the review group who listens to your progress report, the three hundred people at a professional meeting who hear your debut paper, and the half-million television watchers who see your press conference are all audience. The numbers do not define an audience. Their completion of your message does. The members of the audience are your receivers, and they receive you one at a time. Some decode you accurately, liking what you transmit; others succumb to entropy and almost wholly miss what you mean. Keep in mind the singularity of each person's reception, for it is the key to audience understanding.

Especially in the professional world, most of us send our messages out to *multiple* receivers. Rudolph Verderber, in his text on group discussion, says that studies of business practice indicate that nearly 40 percent of business management time is spent in group and consultation contexts.[1] Those multiple receivers present a difficult challenge to someone who seeks to craft messages with minimum entropic loss.

Visualize how complex it is to match the message to different individuals, all of whom have varying backgrounds, yet all of whom are listening to a given communication at the same time. The problem is not unlike that of the "one size fits all" garment. If the wearer happens to be the statistical average, there might be a decent fit. If the wearer is taller, heavier, shorter-legged or longer-armed than the idealized average, the garment will bind or sag. The generic communication aimed at the random generalized audience will not fit any more easily. The "average listener" might find the message suitable; those with unusual attitudes or values in their backgrounds might find it troublesome. Audience analysis is an attempt to improve the odds for the message's fit. The more a communicator discovers about the individuals in the audience, the more likely it is that the message will be tailored specifically for them.

Finding out who the members of a particular audience are and how they think is big business. A poll taker such as A. C. Nielsen, for example, operates in twenty countries on four continents, besides being the major market-research company in the United States.[2] Such opinion takers have recognized the need to find out what consumers want and why they make certain choices for precisely the same reasons any communicator should: this feedback will help structure future messages of improved effectiveness.

Talk of poll takers and market analyzers conjures up pictures of mind manipulation and brainwashing. The image is an unfortunate and probably undeserved one, but stories of unscrupulous advertisers and politicians abound. At times public opinion has been managed without thought to ethics.[3] We presume that you, the students to whom we speak, are persons of integrity with strong values in your backgrounds. If your messages are legitimate ones and you believe they benefit your receivers as well as yourself, analyzing your receiver enhances your message's chance of success.

ANALYZING AUDIENCES

Whether the audience is one person or the entire country, two categories of information about them can help you craft your message:

demographics and attitudes. What follows is a simplified view of how to conduct an analysis so that it yields you sufficient information for constructing future communications.

Demographics

Demography is the science of human-population study. Its task is to measure the number of persons in a given area and then, through comparison with past measurements, to predict future population trends. *Demographics,* a term drawn from this concept, is the study of observably measurable factors in human populations. Such things as gender distribution, educational level, age, occupation, group affiliations, etc., are demographics. Each factor can be documented without reference to a person's feelings or evaluative system. Demographic information categorizes people into verifiable groups such as home-owners versus apartment dwellers, the overweight versus the underweight, parents versus the childless, Democrats versus Independents, etc. The information does not tell you what these people think and feel.

To clarify the concept in your own mind, assemble several friends in a room and categorize them according to ten demographic factors. You could choose:

- Height
- Ethnic affiliation
- Declared major
- Those with corrective lenses
- Those with jobs
- Those from one-parent homes
- Private high school background
- Those who have traveled out of the country
- Religious affiliation
- Transfer students
- Etc.

You must be careful to exclude *inside-of-you* evaluations such as good-looking, intelligent, or pet-lover. You cannot document any of these qualities through observation, other than to say that *you* believe them to be the case. Of course, you could translate these factors into such demographic verifiables as those who have won beauty contests, grade-point averages above 3.5, and numbers of pets owned.

Attitudes

Sometimes, knowing what the receiver's feelings and evaluations are can be helpful. Such knowledge requires going "inside" the other person's head to judge what might be there. One approach is to measure the other's attitudes. Psychologist Arthur Cohen traces the earliest controlled studies of attitude and attitude change to the 1920s and 1930s. At that time investigators were concerned mainly with the degree to which people held certain opinions. They had little interest in understanding the various processes operative in an individual as attitudinal change takes place. Today, however, elaborate and sophisticated studies attempt to trace specific patterns as people react to those message-senders who attempt modification of some response. Cohen remarks, "These experiments [conducted during World War II and later] were among the first to show how specific content transmitted by specific communicators affects particular audiences."[4] The study of attitudes has become a standard component of audience analysis.

Attitude can be defined as enduring, generalized, learned predispositions toward some class of evaluative responses. If you usually (enduring) pick (generalized, whether it is in a restaurant or at home) chocolate ice cream rather than strawberry ice cream (evaluative response), this definition says that you have a positive attitude toward chocolate ice cream as compared with strawberry. Furthermore, this response was learned at some time in your background.[5]

Attitudes are usually divided into affect, cognition, and behavior classifications. The *affective component* refers to a person's liking or emotional response to something. The *cognitive component* is a person's belief or knowledge about that thing. The *behavioral component* concerns the person's actions directed toward that something.[6] Your action with the ice cream can be said to be both affective (you *like* chocolate) and behavioral (you usually *choose* chocolate).

The study and understanding of attitudes is extremely complex, as it must be for any of us to make sense of what is happening inside another. We cannot give you competence in such a study through this book. It is sufficient for you to understand that some measurement as well as inferences based on that measurement are necessary. Two situations apply when you are gathering data about your audience:

(1) you have direct access to the measurement of your audience or (2) that audience is unknown to you.

An Audience That You Know A classroom and a consistent work group are cases where you would have *direct access to measurement.* The members of your audience are available to you before the communication in question takes place, and you can gather data on them. Speakers often use some written form to gather and summarize information about their audiences. We offer Figure 6-1 as one version of this audience analysis form.

Take a look at the summary audience analysis form. Notice that it is divided into three sections: demographics, attitude scale, and adjustments the speaker has to make based on the data. Most of the demographics are general, with the exception of "special pertinent factors." Each topic requires information about the audience that is irrelevant to other topics. If, for example, you were planning to persuade people to install solar-energy devices on their homes, you would need to know how many home-owners you had in the audience.

AUDIENCE ANALYSIS SUMMARY

Numbers of People in Audience ______________ Date ______________
Setting of the Presentation ______________________________

General Topic ______________________________
Action Desired from Audience ______________________________

DEMOGRAPHICS

Age Range ______________ Gender ______________
Educational Level Range ________ Interests ______________
Group Memberships ______________________________

Special Pertinent Features ______________________________

ATTITUDES

How many individuals in the audience would fall into each category below?

Strongly Agree—Agree—Not Sure—Disagree—Strongly Disagree
SA __________ A _____ NS ______ D _______ SD __________

1. I know a lot about this topic.
   SA ____________ A _____ NS ______ D _______ SD ____________
2. I have heard many speeches on this topic.
   SA ____________ A _____ NS ______ D _______ SD ____________
3. I have reached definite conclusions about how I feel about this topic.
   SA ____________ A _____ NS ______ D _______ SD ____________
4. My feelings about this topic are positive.
   SA ____________ A _____ NS ______ D _______ SD ____________
5. My feelings about this topic are negative.
   SA ____________ A _____ NS ______ D _______ SD ____________
6. I am willing to consider new approaches to solving the problems discussed in this topic.
   SA ____________ A _____ NS ______ D _______ SD ____________
7. I believe the problems discussed in this topic are too expensive to solve.
   SA ____________ A _____ NS ______ D _______ SD ____________
8. I believe the problems discussed in this topic are too complicated to solve.
   SA ____________ A _____ NS ______ D _______ SD ____________
9. I believe the problems discussed in this topic are a private matter of decision.
   SA ____________ A _____ NS ______ D _______ SD ____________

SPEAKER ADJUSTMENTS BASED ON FINDINGS

Identify any special difficulties your topic might present to this audience:

____

____

____

Decide how you will overcome any identified difficulties:

____

____

____

Identify additional research needed:

____

____

____

Figure 6-1 A sample audience analysis form.

If, instead, you planned to convince people to stop smoking cigarettes, their home-ownership would be irrelevant, but the number of smokers in the audience would be critical.

The five-point scale in the attitude section allows you to construct a questionnaire based on your topic and then poll your future listeners. Neither validity nor reliability can be claimed for the results based on such uncontrolled conditions. What the experience does teach is that *some* tentative inferences can be made. Rather than give a completely random speech prepared in the absence of any audience data, you now have a chance to control some entropy. Using this method also gets you used to focusing attention on the receiver *before* the communication takes place. That habit carries over into watching what is happening *while* the speech is presented.

It is not possible, nor even advisable, to poll your coworkers in a formal questionnaire. But you can watch and note their reactions as issues come up in casual conversations. People tell you a great deal about their special interests, what viewpoints anger them, and what actions frustrate them. We would guess that people who do pay such attention to others gain, not only raw data for improving their future communications, but also a reputation for being "interesting." Most of us find those people fascinating who pay close attention to us!

Corporations go to considerable effort to obtain data on important future listeners. Standard professional practice makes it mandatory to find out as much as possible about people who will be in the audience. In a very real sense, these persons have been polled. If you are going to appear before a governmental committee or interface with employees of a competing corporation on some cooperative project, your company expects you to know the people to whom you will speak. Many corporations maintain representatives, sometimes in Washington, sometimes in every major city in the country or even in the world. As one of their duties, these representatives provide information about the key people in their areas. They either know your intended receiver or can put you in touch with somebody who does. They can tell you about the person's technical background; major projects, past and present; experience in experimentation or administration; and astuteness. The representative can tell you about "unmentionable subjects" as well as how the receiver plans to use the information you are exchanging. These are serious audience analysis questions, and the professional who walks into a meeting without knowing the answers is just as unprepared as the one who does not understand the test-data summary. Neither deficiency would be tolerated long in the corporate world.

An Audience That You Do Not Know Sometimes, despite all preparation attempts, *the audience is unknown.* When a call comes in January, inviting you to address the March meeting of the Parents and Teachers Organization in an unfamiliar school district, what are your resources? You obviously cannot conduct a formal demographic and attitudinal survey. You may not even have any "representative" who can tell you about the people you will meet. Do you then cross your fingers and walk in without any audience knowledge? We hope you already know our answer. A responsible communicator strives *never* to approach any receiver without some prior information. Even when you stop a stranger on the street to ask directions or time, you have usually done so only *after* the person has been sized up for approachability. For any invited presentation you can do nothing less. The person who contacted you, usually the program coordinator, can provide some crucial audience data. Ask your contact:

- What have the last four or five speakers used as topics?
- Which speaker and topic seemed to receive the best audience responses in the last year?
- Which speaker and topic seemed to receive the worst audience responses in the last year?
- What about the speaker or topic seemed to cause this "best and worst" reaction?
- Is there a neighborhood or township issue that seems "hot" right now?
- What do most of the parents in that school district do for a living?
- Will the audience be allowed to ask questions after the presentation?

Even such minimal information will help you prepare a communication that has a chance of suiting the audience's interests and abilities.

Once you have assembled the best data you can regarding your receivers, keep in mind that the most you can do is reach an approximate fit. The specter of entropy is always present, and generalizations about an individual based on assumptions regarding the group are risky. For example, you could know that most of the people in your audience are college graduates, many of them vote Republican, nearly all have incomes of $50,000 or more per year, and significant numbers of them are members of the Society for the Prevention of Cruelty to Animals. Yet, sitting before you might be an Independent high school dropout, earning $100,000 per year, who abuses the cat. An audience analysis can only improve your probability of controlling entropy with a receiver. No absolute conclusions are possible, and success is never guaranteed.

RECEIVER RESPONSE AND SPEAKER GOALS

If communication is a process of selectively sending certain stimuli in order to have a low-entropy translation response on the part of the receiver, then the focus of the process is receiver response. Too many speakers concentrate on what they themselves like to say and/or do, while ignoring those who will have to decode their messages. In an ideal situation, what the speaker is interested in and excited about matches the receiver's feelings. To explore this ideal, we need to consider speaker goals. In this context, a *goal* is what the speaker seeks to accomplish as a result of the communication—that is, what sort of receiver response is desired.

If we look again at the hypothetical invitation to address the Parents and Teachers Organization, the question of goal arises at once. You were chosen to give the presentation because, say, you are a college mathematics teacher with extensive computer experience. The school district, you discover while doing your audience analysis, is considering the purchase of classroom computers for elementary grades in order to improve students' mathematical competence by grade five. Further, you find that the district is affluent, has used innovative teaching approaches in the past, and has highly educated parents. Your contact tells you that the group would like you to explain how computers might be used with young children, especially as a tool for mathematical mastery. Some speakers might assume that the goal has been identified: The audience is to understand the use of computers in teaching mathematics. Think about that goal for a moment. Is it your primary goal? You simply want the parents to leave your meeting with better understanding than when they came? Probably not. The understanding is part of the goal, to be sure, but you also want them to *approve* the purchase of computers, and/or work with their children to gain ease in numeric concepts via computers, and/or perhaps even volunteer some time to help out in the classroom. You almost certainly want those receivers to think that you are a good speaker and an interesting person too. Can you see how mistaking a goal will cause you to cast the presentation in ways that will miss the mark? Disciplining yourself to ask, "What do I want from this audience?" should help you craft the sort of communication that is likely to get the response you desire.

Traditional Speaker Goals

As long as people have studied communication, they have theorized about speaker goals. Historian Will Durant says that the first set of books we know of was probably the Egyptian *Book of the Dead,* and it was a collection of prayers, charms, and ritual formulae that the faithful recited in order to appease or perhaps even deceive Osiris.[7] Clearly, Osiris was not merely to "understand," but he was also expected to have a specific reaction to the petitioner. In *Rhetoric,* his text on oral discourse, Aristotle defines all communication [rhetoric] as, "the faculty of discovering in the particular case what are the available means of persuasion.[8] Writing for his community, where spoken communication was dominant, Aristotle thus identifies all rhetoric as persuasion. In a modern analysis of the reasons people communicate, David Berlo says, "Our basic purpose in communication is to become an affecting agent, to affect others, our physical environment, and ourselves, to become a determining agent, to have a vote in how things are."[9] Berlo, like Aristotle, sees the goal of communication as persuasion.

Some traditional treatments of communication goals, usually restricted to the case of a speaker addressing at least one other human being, have been broadened and include attempts

- To persuade (to do some action)
- To convince (to think differently)
- To inform (to know something previously unknown)
- To entertain (to enjoy)
- To inspire (to feel uplifted)

We believe these labels mask the underlying purpose of all communication: to alter the receiver. We side with Aristotle and Berlo. In the sense that communication *effects a change* in the receiver, we find it difficult to find any example where persuading another to react was not the goal. Think about it. Your chemistry teacher lectures, seemingly only to provide information, but the result is an actual restructuring of your data about the world. The teacher alters the way you think and, of course, also plans to have you pass tests in order to demonstrate competence. The information is clearly not given for its own sake, but to effect a change in you, yet such speaking traditionally would have been called "informational." When you listen to a comedian or watch a serious drama, these certainly entertain you, but the senders intend a specific alteration in your behavior. The comedian wants you to laugh, while the dramatist and actors want you to weep. It makes sense to us to consider the underlying purpose of communication to be *alteration.* Telling jokes, providing information, and uplifting emotions through stories, and so forth are means to get to that goal.

Contemporary Speaker Goals

Assuming that communication seeks to alter some reaction, we propose three different goals that modern speakers might have:

- Satisfaction of ego
- Motivation
- Action

All of these are invitations to change in both the receiver and the sender. Humor, drama, music, threats, information, and so forth act as agents of change for both the people in the communication.

Satisfaction of Ego Every communication is monitored by the speaker as well as by the audience. It feels good to have receivers respond positively. That goal we cannot ignore. Besides the effect on the listeners, there is also an effect on ourselves. The "I" who is at the center of our lives (egocentrism), must be satisfied with any communication act performed. Even though your supervisor tells you that your report presentation was acceptable or even good, unless you agree, the presentation will trouble you. Most of us work doubly hard when we know that someone special will be in the audience. We want to impress that individual, surely, but we also want to satisfy ourselves that the job is one we think the others will admire.

Motivation By *motivation* we mean changes the sender hopes to make in the thinking and knowledge of the receiver. In this sense, we have taken the definition of motives used by psychologists and applied it to what happens in the minds of people. In their introductory psychology text, Houston, Bee, and Rimm define *motives* as, "(1) what activate and arouse the organism, and (2) what direct the organism's behavior toward the attainment of some goal."[10]

Motivation is, in part, a language function. As a speaker places new ideas, which are at least partly linguistic symbol systems, in the thinking of a receiver, these restructured thought patterns sometimes result in motivation to action. You may not have known much about hunger in America, for example, until you heard and saw a slide presentation on the plight of street people in our cities. The words and images made you aware of something that had not been part of your thinking. At some later time, this restructured pattern may have motivated you to contribute time or money to a center-city hospice. Your kindergarten teacher showed you how to trace letters of the alphabet or how to tell time so that you would *do* these things as a functioning member of the community. The teacher also wanted you to reconstruct your thinking with that information. Knowing how to write makes it likely that a person will choose to write. Knowing how to translate a clock's symbols changes the way a person acts about being somewhere at some time. Written language and concepts of time alter a child's view of the world, and life is never quite the same again.

Sometimes, hearing some piece of information does not cause you to do anything, but it most certainly alters what you carry around in your mind. Perhaps all changes in thinking eventually result in action. For our purposes, realize that the speaker renders the receiver different by altering the thinking of that receiver. This change of mind is what we mean by the goal of motivation.

Action Often a speaker's goal is to activate directly. We communicate because we want some specific change in action to take place. When you ask a passing stranger for the time, and the stranger says, "It's 3:15," you have caused a behavioral alteration in that receiver. After all, the stranger's life was interrupted, and the overt action of looking at a watch occurred in response to your communication. Salespeople speak to get your signature on a contract or the money from your pocket. Whether they lead companies or countries, presidents speak so that their listeners will not support the actions of their opponents. You speak, sometimes, so that the person who attracts you agrees to meet somewhere at some prearranged time. You respond to interview questions in ways you hope will cause the interviewer to make a job offer to you rather than to another candidate. In each instance, you might use humor and will almost certainly use information. Neither provoking a laugh nor providing information is the aim of your message. You use them *as means* to secure the actional response desired.

Most communicators have an immediate speaking goal and an eventual goal. A politician addressing a group of potential supporters does not expect the audience to vote immediately. The election might be many months away. The *immediate* goal might be to have the listeners' feelings of neutrality altered to "What a capable leader this person would make." The *eventual* goal is to translate those positive feelings into the actual pulling of the voting machine lever. An immediate goal for any large audience speech might be motivational: *to change an audience's thinking.* It might also seek an action: *to applaud in support of the speech.* The eventual goal is probably always an action: do something in response to the speech topic. The politician, after all, would be extremely upset if voters carried away positive feelings after the speech but later cast their ballots for the opponent.

Using the framework of immediate and eventual goals, think about the following communication encounters:

On Campus:

- *Setting:* A speech class during which you deliver your final presentation.
- *Immediate Goal:* Students will understand the differences between passive and active solar-energy applications for domestic buildings.
- *Eventual Goal:* Students will choose passive solar methods of building when it comes time to construct a house.

  Teacher will give this presentation a grade of A.
- *Setting:* A lecture hall where an astronomy professor teaches a lesson on the stellar spectra.
- *Immediate Goal:* Students will understand the differences between continuous, bright-line, and dark-line spectra.
- *Eventual Goal:* On the midsemester exam, students will accurately answer the question about spectroscopy.

- *Setting:* Campus coffee shop where you tell funny stories about your roommate to three friends.
- *Immediate Goal:* Friends will relax, laugh at your stories, and feel sorry for your situation.
- *Eventual Goal:* Friends will continue to treat you as an "all-right person."

In Professional Corporations:

- *Setting:* A quiet restaurant where you and a prospective customer are having lunch.
- *Immediate Goal:* Customer and you will relax, forgetting the preliminary test in progress.
- *Eventual Goal:* Customer will assess you as a socially adept and acceptable person, choosing you as the test liaison.
- *Setting:* The test chamber during the tour that you conduct with four prospective customers.
- *Immediate Goal:* Customers will understand the equipment used and the daily schedules followed during testing.
- *Eventual Goal:* Customers will recommend that your company be chosen as the site of the upcoming test.
- *Setting:* Your meeting with your manager in order to conduct your yearly work-performance review.
- *Immediate Goal:* You will say nothing that jeopardizes the 20 percent raise you anticipate.
- *Eventual Goal:* You will maintain a "comer" reputation that results in a promotion.

Can you see how thinking about goals, immediate and eventual, casts the communications into an entirely different light? The speaker determines what the goals might be. The speaker's goals and the receiver's goals might also vary, even though both are engaging in the "same" communication. In the last example, after all, the manager might simply want to get an unpleasant task out of the way as quickly as possible, because the manager's eventual goal could be to remove you as a threat to his or her own advancement.

It is often uncomfortable to think about our various encounters as goal-oriented. Ideas of manipulation and lack of spontaneity arise. Our position is that communication always involves alterations. We misunderstand the process if we ignore the reality that, when we reach out to another, we alter that other person's activity. Whether the reaction is one of egocentric satisfaction, in-the-head motivation, or overt behavioral action, each of us changes a present state into something else.

GOALS IN THE PROFESSIONAL WORLD

The modern corporate world is extremely goal-centered. Its members, technical or not, are expected to set overt objectives and demonstrate that they have met them. Communications, both inside the company and beyond, can be thought to have four purposes, all centering on the eventual goals of either motivation or action. All of them also can provide a great deal of egocentric satisfaction. We maintain, however, that the final corporate goal is to sell goods or services, an act that is the company's reason for existence.

Acceptance of Image

Some corporate communications are transmitted in order to establish the company, its purpose, its products or some individual within it as worthy of belief. The degree to which your company commands belief, the reputation its representatives carry into a communication, is called *credibility*. It is related to "Your Version of Me" in Figure 5-1. Aristotle thought that credibility, which he called *ethos*, was the most potent of all the tools that a speaker had to affect persuasion. Personal and corporate image making in order to improve credibility is still regarded as vital.

Corporate Image Making Company and product reputation are, like the individual's, based on background. Record of the company's past success, integrity of personnel, ability to deliver within deadline and budget are all part of that background. Knowledge about such past practices works its way into the background of the customer as well as the speaker. Expensive advertising, elaborate public relations, and marketing departments work to protect a company's positive image.

The technical staff is part of that protection too. No professional works in a credibility vacuum. Much time is spent, usually by technical professionals, enhancing the technical reputation of a corporation. Potential contract customers are invited to observe some work in progress. They tour, watch the personnel at daily tasks, examine the specialized equipment, and have an opportunity to discuss details with the staff who will perform the job. Such tour-guide duty is extremely common and certainly time-consuming. Senior staff members almost always draw the task. The more important the customer, the more senior the tour guide is; and both customer and guide are more than likely to have technical training. A company's willingness to commit a highly paid individual's time to such public relations duty is ample evidence of the importance of credibility building.

Often the reputation of a company pays off in contract dollars. Private and government agencies sometimes invite certain companies to submit proposals for upcoming projects. Without an existing image, a corporation has a difficult time even getting to the bidding stage of contract.

When the corporate reputation for certain work is exceptionally strong, customers might be willing to pay a premium. Working on the theory, "The higher cost is really a better value," contracts are sometimes awarded to those in that enviable position. Such an assumption is really no different from what you do when you purchase something from a given store at a higher price than is advertised at another store. You recall that the first store treated you well when you had a problem with a past purchase or that its sales staff took the time to help you make a choice. If you also recall a discourteous clerk from the second store who did not seem to know much about the stock, your investment of a few extra dollars to receive better value does not seem extravagant.

Personal Image Making Personal reputation is an individual's version of corporate credibility. In this book we propose to make you a competent communicator worthy of a respectful hearing. Our position is that you reveal yourself in all your communicative acts. The version of you decoded by others, even though you are not "really like that," forms your reputation. The more positive the version, the higher is your credibility.

Personal reputation within a company is extremely important. A person might gain credibility simply by being associated with some task-effective group. Receivers assume that the group would not accept anyone who was not also effective. This reflected-glory or guilt-by-association effect can work in the corporate world just as unfairly as it does in classrooms. Some of you learned to your dismay that your parents' advice about "bad company" was accurate. People did assume that you were exactly like those in the "bad" group.

Young professionals who have worked with disastrous projects also have a great deal of credibility-building to do before the association fades.

Being part of a successful group is a major career boost. Often, even after a promising project is abandoned for lack of funding, the personnel who worked on the project are very much in demand. Colleges lend a portion of their reputation to their graduates. Some companies will not interview people from certain degree programs; others give preference to those who graduate from a given place. The esteemed mathematical program in the Courant Institute of the [then] City College of New York, for example, enhanced the individual reputations of its graduates and those who taught there. Such associations are important, for they give the young professional the first chance at personal credibility.

Once hired and working within a group, your credibility can grow or decline. A reputation is a fragile thing. You may have told a single lie in your entire life; but, if you were caught at it, you gain the reputation of liar. Unfortunately, the opposite is true also. You may be a habitual liar; but, if receivers think you are honest, your reputation is that of the truth-speaker.

Professionals need to be mindful of their credibility. If you develop the image of being difficult to work with, unreliable in times of stress, uncommunicative about processes, unable to deal easily with contacts, and so on, your career will be in jeopardy. Look to your communications if you suspect that a negative impression of you is afloat. *Some* of your messages are obviously getting lost in entropy. Nobody starts out wanting to be decoded as unacceptable. Identifying what you might be doing to cause such a mistranslated message could help you make changes.

Acceptance of Ideas

In the mid-1960s Thomas Kuhn, a theoretical physicist, wrote a remarkable book called *The Structure of Scientific Revolutions*. In it he argued an unusual thesis regarding how new ideas in science come to be accepted. The history of science shows that discoveries are not simply single additions to the knowledge level of communities. They are most certainly not the result of the calm, objective examination of facts that often characterizes the stereotypes people have of scientific practitioners. Discoveries are controversial and sometimes hotly debated matters of scientific *acceptance:*

For these men [the specialists in the area of new discovery] the new theory implies a change in the rules governing the prior practice of normal science. Inevitably, therefore, it reflects upon much scientific work they have already successfully completed. That is why a new theory, however special its range of application, is seldom or never just an increment to what is already known. Its assimilation requires the reconstruction of prior theory and the re-evaluation of prior facts, an intrinsically revolutionary process that is seldom completed by a single man and never overnight.[11]

Kuhn's book traces some of the most spectacular scientific theories and how they were received. Landmark classics such as Aristotle's *Physica*, Ptolemy's *Almagest*, Newton's *Principia* and *Opticks*, Franklin's *Electricity*, Lavoisier's *Chemistry*, and Lyell's *Geology*, all superseded the ideas prevalent in their day. These theories, in turn, gave way to others evolved by later scientists. Neither the scientists nor their theories were accepted merely through a cool reasoned examination of facts. Each new theory involved a revolution, and not until the disciples of the revolution won power was the theory accepted.

In some cases, the guardians of the old ideas resorted to falsifying data, physical torture, and coercion more usually associated with political and social upheaval. In his introductory statement to the work of Galileo Galilei, Dagobert Runes describes how Galileo was punished for his publication of *Dialogo Sopra I Due Massimi Sistemi Del Mondo*. He was incarcerated, tried by the Inquisition, forced to recant his findings, and sentenced to prison.[12] Kuhn claims that the scientific community, at many critical discovery points, encouraged such suppression of new ideas.

Kuhn's thesis reminds us that all technical advancement involves the acceptance of new ideas. Science, after all, is practiced by human beings, and they must be convinced before they will accept. Alfred North Whitehead, commenting on the personal cost to Galileo of publication of his work, says, "It was paid for also in the heart's blood of the author."[13] Gaining acceptance involves motivational change and then action. Scientists, like other human idea-shapers, must use persuasive communication for novel ideas.

Much of a technical corporation's communication centers on the means of selling its own ideas as better than those of competitors. New approaches come under intense scrutiny. Does the innovation under consideration do all that the old method did? Does it do so with some inherent advantages? Does it warrant the gamble change involves? Will it be cost-effective to implement? If the idea is not accepted, it dies in the laboratory. Whether it becomes accepted is, in part, the responsibility of the scientists who send and receive messages about it.

Acceptance of Methodology

Closely related to the acceptance of ideas is acceptance of the methods through which the superiority of those ideas will be demonstrated. To the technical professional, demonstration must involve experimentation, and the methods of that experimentation must be credible. Examine these five key experimental expectations.

1. Have Measurements Been Made Using Both the Old and the New Method? If so, does the new have demonstrated advantage over the old? Temperatures are sometimes measured, for example, on hand-held optical pyrometers or with infrared thermometry, rather than by the established method of using direct contact thermocouples. Comparisons have to be made using both methods and acceptable reliability levels have to be reached before the new instrumentation readings will be used. Beyond that, it must be established that the instruments have the inherent advantage of being indirect: they can take measurements at a distance, picking up hot-spots through heat leaks from engines or houses, for example. If such procedures are seen as acceptable, the new measurements can be used.

2. Is the Quantification Accurate and Complete? Scientific data are almost always reduced to numeric concepts. Mathematical constructs used must be accepted as pertinent and properly applied. If statistical extrapolation is employed, each step must be justified as correct. Computers have simplified the handling of some of the large numbers, but the very presence of computers has probably increased the expectation for quantification. Be prepared to deal with growing demands for numeric backing.

3. Have Enough Experiments Been Run? There is no magic number to say when "enough" is *enough,* but some satisfactory point does exist. Perhaps it seems sufficient when the experiments have covered a range of operations. If the apparatus is accurate for a range of 200 degrees to 1200 degrees, that fact may be convincing. Perhaps it takes a spectrum of conditions. If the chemical protects nine different metals in different weather conditions for a period of a year, that observation may be convincing. It is part of the communicator's duty to determine what kind of substantiating data are needed to gain acceptance for the idea proposed.

4. Have Unexpected Results Been Explained? Accounting for "limitations of techniques" is extremely important. When the test suddenly demonstrates a voltage output higher than predicted, does this invalidate the principle under study? It might, unless the surge has some reasonable cause, and the cause does no harm to the process being verified. Unusual test results cannot be ignored, for they might be significant. Unreported exceptions to the rule and surprises have no place in scientific investigations. Their presence must be articulated carefully or else they might destroy the credibility of the entire experiment.

5. Is There Willingness to Adjust Expectations? The surprises in the test results might not mean that the experimental object was a failure. It might indicate, instead, that refinement is needed. That piece of equipment that proved reliable in a heat range of 200 degrees to 1200 degrees might have been meant for conditions of 1500 degrees. The test demonstrated that it could not work under that amount of heat, but it might have applications for use within its 1000-degree range.

At each step, the operations of the scientist are communication decisions. The "how" of a demonstration is a matter of matching method with audience expectations. The goal is always acceptance, and the credibility of the ideas proposed and the methods by which they will be demonstrated must be seen as legitimate.

Acceptance of Contract

The culmination of all this communication is the securing of a contract. Corporations are in the business of making profits and/or providing services. In modern business, this involves contractual agreements. For a technically based profit-making business, contracts cover all or any portion of the following:

- Feasibility study
- Developmental exploration
- Conceptual design
- Developmental testing

- Preliminary design
- Engineering testing
- Final design
- Final testing
- Delivery of hardware/software

The process begins with a "bid package," the customer's definition of how broad or narrow the job is to be. There is often a competitive component, which involves the customer's inviting proposals from a number of companies to bid on any or all of the contracts, as limited by the bid package. Sometimes contracts are awarded to multiple winners in the early stages so that a final design tested through the competitive process emerges.

A company then decides on whether to bid for the contract. The bid package might be one small phase of the entire test and design, or it could be some substantial part. It could also be for the whole project. In any case, the steps are similar. The customer submits the specifications for the scope of the bid. In response, the corporation indicates its intention to submit a proposal. The proposal includes, besides the technical information dealing with process and method, a defense of effort levels needed to complete the contract. Personnel hours, equipment, schedule planning, delivery dates, and all corresponding budget information must be included. These steps take limitless hours of meetings, clarification phone calls, in-house information exchanges, audience analysis of the customer, on-site visits, in addition to development of the technical process, which must be under way. Each step involves intense communication, usually oral, and the process culminates with the submission of the proposal.

If the proposal wins, the final contract is signed and work begins to comply with the agreements reached. At this point, usually only three options remain: bring the contract in within deadline, secure an extension, or change the scope of the contract.

Testing and design changes make the second two options a decided possibility. Schedules do slip, sometimes beyond the control of the customer or the corporation. Labor or transportation strikes, inability to secure needed equipment, illness among key personnel, or any of a thousand gremlins might make it impossible to meet deadlines. Such cases place the corporation in serious credibility threat. Careful communications are needed to insure that reputation is not damaged. Customers tend to be wary of any excuse, so backup plans are often part of in-house proposal planning.

If the schedule must be slipped, it can be extended on a no-cost (no additional monies from the customer) or a cost-plus (additional monies paid by the customer) basis. These options are usually articulated within the original contract.

If the customer or the corporation sees the need to change the scope of the contract level, this need usually arises while the study is under way. Testing may indicate that the problem is far more complex than anyone anticipated. The customer might decide that additional refinements are needed, and these extend beyond the original agreement. By mutual consent, and under the firm guidance of legal counsel, a contract can be altered in scope. The original Mercury space capsule, for example, was to be built and tested for about sixty million dollars. As the problems of space travel unfolded during testing, the contracts were altered significantly; and the final price tag was far higher than the original cost.

At each step in a crucial contract process, communication decisions are made. How something is written, whether a phone call or an in-person meeting takes place, how personnel answer important questions during review meetings, whether there is accurate in-house information, whether the lead personnel hold together the morale of their staff are all matters of people talking to other people. Done well, the contract is secured and completed. Done badly, the individual and the corporation suffer.

WRAP-UP

When we describe the steps that culminate in a contract we are actually describing the process of moving from self to others. It is a process inherently similar to what happens when any message is sent forth to reach some goal. We began this chapter with a concept of audience. That customer is audience, and so is any receiver. In corporations, all the resources a company has are brought to bear in finding the best information about the audience. So ought it to be in any communication.

Business goals are usually easy to identify: get the signature on the contract. Personal goals in daily encounters are often not so obvious. If you fail to consider all the communication strategies that lead to resultant action, you will misunderstand the entire process. If you fail to see its parallel in all daily communications, you will miss the focus of human encounters.

The customer (or friend) is informed, inspired, and finally persuaded. The lunches (or dates), during which personal credibility is established, the tours (or classroom recitations), during which technical reputation is enhanced, the discussions about testing methods (or late-night arguments with a loved one) that motivate an acceptance of approach, all result in the final decision to reach or fail to reach some goal. They are all part of the motivating or/and activating process called communication. Those who enter the technological world accept the responsibility of communicating effectively. Remember who decides if you achieve your goals. It is the audience, that person who waits for you at the receiving end of your communications.

ACTIVITIES

1. Using your class or some other group to which you belong, make a list of the demographic factors you might be able to measure.
2. Select an attractive advertisement for some nationally advertised product. In a three-to-four minute presentation to the class, explain:

    - What audience analysis assumptions you think the product's manufacturers made before they produced it.
    - What audience analysis assumptions you think the product's advertisers made before they designed the ad you are using.
    - How, if at all, the two analyses differ.

3. Assume that you are a student campus leader who has an appointment with your university president because you want reversal of some policy students oppose. What would you need to know about your president to help you prepare for that meeting? How would that knowledge help you restructure what you plan to say?
4. Select some attitude you hold on some controversial subject. Make a list of what a communicator would have to do to alter that attitude for you, even if the alteration is small. How might that communicator find out about what you would demand before you would be willing to change your attitude?

5. Attend a meeting featuring a speaker who draws an audience where people differ in age, education, interests, and backgrounds. A church service, politician's "open meeting," or a community/campus lecture series would be good examples. Note:
   a. How does the speaker adjust the presentation to this "generic" audience?
   b. How successful were the adjustments?
   c. Which portions of the audience seemed left out of the presentation at which point?

NOTES

1. Rudolph F. Verderber, *Working Together* (Belmont, Calif.: Wadsworth, 1982), 5.

2. Herbert I. Schiller, *The Mind Managers* (Boston: Beacon Press, 1973), 138.

3. See, for example, Schiller, *Mind;* Leonard W. Doob, *Public Opinion and Propaganda* (Hamden, Conn.: Archon Books, 1966); and Alfred McClung Lee and Elizabeth Briant Lee, *The Fine Art of Propaganda* (San Francisco: International Society for General Semantics, 1979).

4. Arthur R. Cohen, *Attitude Change and Social Influence* (New York: Basic Books, 1964), 1.

5. Philip Zimbardo and Ebbe B. Ebbesen, *Influencing Attitudes and Changing Behavior* (Reading, Mass.: Addison-Wesley, 1969), 6–7.

6. Zimbardo and Ebbesen, *Influencing*, 7.

7. Will Durant, *The Story of Civilization: Our Oriental Heritage* (New York: Simon and Schuster, 1954), 202–3.

8. *The Rhetoric of Aristotle*, trans. Lane Cooper (New York: Appleton-Century-Crofts, 1932), 1.2.

9. David K. Berlo, *The Process of Communication* (New York: Holt, Rinehart & Winston, 1960), 11–12.

10. John P. Houston, Helen Bee, and David C. Rimm, *Invitation to Psychology*, 2nd ed. (New York: Academic Press, 1983), 312.

11. Thomas Kuhn, *The Structure of Scientific Revolutions*, 2nd ed. (Chicago: Univ. of Chicago Press, 1970), 7.

12. Dagobert D. Runes, "Galileo Galilei," in *A Treasury of World Science* (Paterson, N.J.: Littlefield, Adams and Company, 1962), 324–55.

13. Alfred North Whitehead, *Science and Philosophy* (New York: The Wisdom Library, 1948), 239.

CHAPTER SEVEN

RESEARCH AND ORGANIZATION OF PRESENTATIONS

An oral communication is a transient thing. It lasts a few seconds or several hours; but, when it finishes, it is gone. Books, reports, and papers can be picked up and experienced time and time again; speech cannot, except as a memory in the mind of the receiver. Even if an oral presentation is tape-recorded or videotaped, the machine version—a flat-surfaced image with a voice distorted through recording devices—is inherently different from the live event. The nature of oral discourse is transience and that is perhaps the main reason for its heavy entropy.

Two unfortunate assumptions mislead many who want to improve their oral communication:

Error One:

- Preparing for speaking is the same as preparing for writing.

Error Two:

- Speech is widespread and "natural," so it needs no preparation.

The first assumption involves a half-truth, a kernel of accuracy. Both speaking and writing involve audience analysis, research, organization, choice of symbols, and presentation of those symbols. For the writer, the task finishes when the manuscript goes through final editing. For the speaker, the communication has not yet begun. After completing the research, organization, and arrangement decisions, the speaker comes before a receiver and makes the speech come alive through enactment. Careful preparation increases the probability of competent communication but does not guarantee it. Between the magnificent plans and the blundering mind block lie many very upset speakers.

The second conclusion is untrue. We are certainly part of what has been called the "Communication Society."[1] Just because oral exchanges are so common, broadcast widely by radio, telephone, and television, we cannot assume they are done competently. The briefest examination of any daily newspaper shows stories of people misunderstanding, acting on these misunderstandings, and involving themselves (and sometimes their countries) in tragic consequences. Most of these actions are not errors of reading and writing, but those of speaking and listening.

For the technical professional, as for every opinion maker, the need to prepare before the speech is paramount. In this chapter, we will discuss how you take the results of audience analysis and match them to a topic. Once you select the topic, you research and refine it. Finally, you organize the information into the best possible pattern for achieving your goal. Finally, you have a chance for a successful speech.

AUDIENCE ANALYSIS AND TOPIC SELECTION

The choice of topic in the business world is tied to the speaker's position within a company, which predetermines certain recognized subjects of expertise. But topic selection is not always so easy. So that you can understand the differences between a relative lack of choice in topic and nearly unlimited choice, we will consider each separately.

Mandated Topics

Some topics are handed to you. You may not have chosen them; you may even have little interest in them, but they nevertheless become your assignment. This is a difficult situation in the classroom. What student has not groaned inwardly when the teacher announced, "Each of you will research some aspect of hazardous-waste disposal, and prepare a five-minute speech (or ten-page paper or forty-minute panel) on the topic"? You have little recourse. You prepare the assignment or drop the class.

In corporations your choices are often no wider, but it is far less likely that you will be completely uninterested. After all, you know where your department fits into the contract proposal. You probably were hired into that department because of your special knowledge or training. When you discover that you have been assigned to present the progress report on your testing during the weekly review meetings, you already have a vested interest in your section's work.

Moreover, whether you are or are not brimming over with enthusiasm for your topic, the responsibility for preparing and delivering it is yours. Fortunately, with research and knowledge, people do warm to their subjects. Perhaps it is a clever way our systems have of allowing us to adapt to situations, but understanding often does bring interest.

Personal-Choice Topics

When the topic is of your own choosing, selection might be a problem. Whether you are invited to talk about "anything" dealing with energy as the invited speaker to the Rotary Club, or your teacher tells you to pick "some topic" for your final presentation, you are going to be frustrated for a while. The following steps can help cut down the selection time:

- Start with what you know or what interests you. Your audience analysis should tell you if this is the subject your listeners also want to hear about.
- Consider some problem you yourself might have. If it is a difficulty for you, perhaps it is widespread and could stand investigation.
- List the issues that seem "hot" at the moment in your town, company, or campus. The issues might form part of a wider problem.
- See what the daily papers highlight. Does any of the local, state, national, or international news suggest something?
- Check the *covers* of the newsmagazines for the last few months. They tell you what topics have been important enough to highlight in cover stories.
- Review the papers you did for high school, for other courses, or for other departments. Is there something there that might be updated and worked into a presentation?
- Have a long intrapersonal talk with yourself. Your special skills, experiences, background, or knowledge could be refined into a topic.

Each of the preceding ideas has worked for someone. They are a good beginning point for you. They are not, however, the steps to your final decision. For that, the audience analysis is necessary. If you discover that nobody wants to hear about the relative merits of internal combustion versus turbine engines, you have a problem. You might find a way to make that topic interesting. If not, choose another. If you have discovered that your audience is sick of hearing about uniform drinking laws, you might be able to treat the topic in a fresh and novel manner. If not, choose another. These subjects may be extremely interesting to you, but you are not the sole receiver of the message. Grow accustomed to thinking in terms of "What do I want from this audience?" and you will feel comfortable matching your choice of topic to the needs of your receiver.

TOPIC RESEARCH

A good oral presentation begins a long time before the first word is said. It makes a pleasant myth to visualize Abraham Lincoln jotting down a few notes on the back of an envelope on his way to Gettysburg, or, say, William Jennings Bryan breaking the deadlock of the stalled 1896 Democratic convention with his spontaneous "Cross of Gold" speech and emerging as the convention's choice. Lincoln's and Bryan's friends and colleagues embellished these very stories, but they are untrue. Lincoln wrote several drafts of the Gettysburg Address, some many weeks before he spoke it. Bryan actually had given his stirring speech, in various versions, several times before it came so "spontaneously" in the convention hall.[2] So much for myth.

Most of us are more acutely aware of the preparation that goes into presidential speaking today than were the people in Lincoln's or Bryan's day. The press and best-seller books tell us about small armies of researchers, speech writers, and delivery coaches who prepare candidates for a single important speech or televised debate. When you consider the extensive experience and training many candidates have already had, as lawyers, business executives, governors, or actors, you can see that there are no shortcuts to preparation.

Research is a vital part of preparation. For the student and the technical professional, some predictable sources of information exist. We shall consider four: written, interview, personal, and consultant.

Written Materials

As a source of material for your topic, you probably think of information in print first. For most of us, our training in research began with a parent or teacher saying, "Look it up." The advice is good; but some subdivisions of written sources especially helpful to the technical professional exist.

Journals and State-of-the-Science Resources Professional organizations publish research papers and test data in their journals. These are especially helpful because they often present theoretical ideas just being introduced to the scientific community. The place for new ideas or challenges to established ones has always been the pages of professional journals. Each discipline has several organizations, and each organization has at least one journal. Student membership, usually offered at attractive discount, will give you a subscription to the journal you are most interested in. Your university library should have the rest. Corporate libraries subscribe to them routinely. Popular publications, such as *Scientific American, National Geographic,* or *Byte* provide another type of help. They are not usually as thorough as the professional journals, but they have interesting and easy-to-understand articles within some very specialized areas. They also tell you what theory has advanced beyond the journal audience into the wider community.

State-of-the-science materials detail the hardware available to a discipline. They consist of manufacturers' specification sheets, laboratory books, and data sheets. In many ways they function as catalogs for equipment. Disciplines change rapidly, so lists of available chemicals, batteries, and test materials are as important to students as they are to corporate professionals.

Books Books take a long time to write and distribute. In a rapidly changing discipline, the lapse of time between an author's conclusions and publication may render the book out-of-date as soon as it is printed. If the book is a text meant as a teaching device, the time gap may be unimportant. In the professional world, however, current information may be paramount. A person doing technical work for a corporation is assumed to already have basic knowledge. What is usually needed for a presentation is the additional knowledge that latest research brings. Information time lags may be a serious handicap if you rely exclusively on books.

Reference books, on the other hand, serve a valuable function for both students and professionals. They are repositories of needed facts and are not so subject to obsolescence as are other types of books. Sources such as the *Handbook of Chemistry and Physics* compile their data and facts in useful form; as summaries, they are valuable effort-savers and are widely used.

Government-Sponsored Research and Hearings Governmental agencies and research groups are usually required to present final reports of their findings. Transcripts of hearings on new or controversial issues are usually available to interested persons. Libraries order some government publications of a popular nature as a matter of course. Become familiar with the government documents portion of the library. It is sometimes an underused resource for recent information. Specialized reports are unlikely to be available in most libraries, but these can be requested or ordered privately from the National Technical Information Service (NTIS) of the United States Department of Commerce.

Overlooked Sources At least two sources for technical information are often overlooked: convention papers and on-campus or in-house research.

Most professional organizations convene at least once a year for the purpose of hearing papers of interest to their members. Such papers are almost always selected through a competitive process and tend to be at the forefront of emerging theory. Their authors use the forum of the convention to garner reactions and recognition. Well-received papers may later be reprinted in the organization's journal. Since the convention hears the papers nearly a year before they can be processed and published in those journals, the information is extremely recent. You can often secure copies of papers by writing to the authors directly through the professional organization. In addition, organizations will usually respond to requests for the printed conference program where titles of individual presentations and names of authors appear.

Frequently, someone on a campus or in a corporation is doing research that has not yet reached the stage of publication. A university biologist might be working on a project that you could examine. A corporate colleague might be developing a mathematical model of considerable help to you. Do not neglect to check out such sources.

Interview Materials

At least three different interview sources are available before the speaker has to resort to the use of paid consultants. Two of them are available to students, and it is worthwhile to investigate these. First, on campuses, students might be able to secure the opinions and guidance of teachers and administrators. Most teachers are more than willing to discuss their areas of expertise with interested students.

Second, you might be able to secure meetings with in-business professionals. These people, like teachers, are often generous with their time when the request for information comes from a young person investigating some project of interest to the expert.

For corporate professionals, discussions with peers in other departments or other branches of the company are possible. For the price of a lunch or an "I owe you one," good information might be readily available. Such exchanges of knowledge are usually reciprocal and, indeed, form the basis of corporate networking. The information-giver will probably feel free to call on you for data in your expertise.

Personal Experiences

Events that have happened to you or those around you might be valuable sources of research information. The professional who describes the successes or failures of some past test and the student who outlines difficulties in the chemistry supply lab can pass on valuable lessons to others. Such experience is often the first clue that something might be wrong (or right) with a theory or procedure. So long as an investigator does not assume that, because it happened once, it is necessarily significant, the retelling of personal stories is an excellent starting point.

When personal stories happened to someone else, they are usually called case studies. When researchers examine a number of such cases and find similar trends in each, they begin to have confidence in the reliability of the findings. Such case studies are equally important to students and to professionals.

Experiments are also personal experiences in the sense that they are a record of what someone did and found under carefully controlled conditions. Technical-subject researchers may look to test results as a rich resource.

Consultants

Consultants are people hired for their expertise in order to give information and/or advice. Corporations frequently use such resources. Those professors, department administrators, and in-business professionals who, in a generous donation of time, might be willing to talk to students, are usually paid well to talk to corporations.

Consultancy firms organized around some specialty provide research services and/or advice for hire. Individuals with uncommon knowledge also sell their time to those who need their expertise. Whether an entire firm or a single person, consultants are valuable and timesaving resources. Their services are not inexpensive, for they usually work on a fee-plus-costs basis. They are hired because they can readily give some information that others do not have and cannot get easily. When a test is in jeopardy or a product's development is stalled, their service might be an extremely economical move.

TOPIC ORGANIZATION

When you have examined all the sources we have discussed and your research is complete, you are now ready to organize the presentation. The research results are like the raw materials from which you build a structure. The cement, stone, brick, and wood can be combined in any number of ways. Some might yield a beautiful and warm house; some might make a pleasing and efficient office complex; some might be plain awful, fit only for demolition. The same is true of the raw data—background materials from written sources, interpretations from interviews, and personal experiences. Put together one way, the materials form a minimum-entropy piece suited to a receiver. Put together another way, the materials put the same receiver to sleep or cause utter confusion. How you construct a presentation is a matter for serious consideration.

Assuming that you already know your audience, are aware of your specific and eventual goals, and have completed thorough research, your organizational pattern must accomplish at least these five objectives:

- Make listening easy enough for minimum-entropy decoding.
- Provide a structure for comprehensibility so that main points are identifiable and memorable.
- Direct the listener clearly to the most important segment(s) of the presentation.
- Satisfy the listener's requirements for method and evidence.
- Enhance the probability of achieving the speaker's goal(s).

No matter how successful a particular topic's organization was for some other audience, if you cannot achieve at least these five ends with the audience for which you now prepare, do not use the same organization.

Basically, any oral presentation has only three main parts: a beginning, a middle (or body), and an end. You doubtlessly realize, however, that we leave a great deal out when we refer to three parts. The variations in those beginnings, middles, and ends are the difference between Martin Luther King's "I Have a Dream" speech and the frequently boring high school valedictory speech. We are going to discuss the purposes beginnings and endings serve and then explore some development patterns for the middle.

Beginnings and Endings

The first few lines you say to an audience ought to accomplish something more than demonstration of a workable voice. The beginning of your speech helps shape listener response. It allows the listener to

- Get acquainted with you
- Become accustomed to your voice and manner
- Establish your credibility
- Focus attention on you and your topic
- Keep listening

These responses are not trivial. Each is important to your eventual goal realization. Speeches that begin, "I'm going to talk about strip mining," accomplish little more than a ho-hum reaction. Work for the audience's willingness to listen to you. If you allow receivers to succumb to the entropy of their own thoughts rather than attend to your message, you might as well sit down.

Speech endings are more than a graceful way of getting the speaker from the podium to the seat. They are your last chance to

- Solidify high speaker credibility
- Summarize what you consider most important
- Clarify the audience action you seek
- Give the audience something to remember

Think of all the speeches you have heard that ended, "That's all I have to say. Thank you very much. Are there any questions?" Not a very memorable ending, is it? Neither does it help you recall what the message and purpose of that speech might have been. What a waste!

Body of the Speech

The middle of the speech is usually called the body. It is the heart of the presentation, the place where each point is introduced, explained, supported, argued, and linked. You can structure the body of your speech in a number of ways. We are presenting a few of the patterns that technical professionals might find most helpful.

Chronological Organization Your research materials could be arranged as they occurred according to time. Using such an early to latest (or most recent to most antique) pattern is appropriate for some subjects. Science has an extremely dramatic history. A historical account of scientific ideas about the elements, for example, might be a way for a chemist to show the evolution of acceptance for a chemical process (and the need to fund continuing research).

Spatial Organization Some topics might benefit from a presentation that arranges information according to physical space. The *where,* rather than the *when,* would be relevant with subjects such as coal or oil deposits, or an argument for increased strip mining. Differences in geographical locations are central to arguments for (or against) programs of wind energy or geothermal development.

Pro-Con Development Some experiments done in the 1950s indicate that attempts at persuasion are most effective when two sides of an issue are presented rather than one.[3] This two-sided presentation, showing the positive side of an issue and its counterarguments, works best when the receivers are

- fairly well educated
- likely to hear or to have already heard opposing views
- already somewhat committed to the opposite side

With these listeners, you should present your side and then some of the arguments from the opposition. As you answer these opposition arguments, your chances for receiver acceptance seem to improve. Assuming that your audience fits the description, this pattern might work in topics such as the health risks of saccharine versus sugar or whether or not caffeine poses a health problem. Rarely will the audiences you face be convinced by hearing only one side of such topics.

General to Specific In this pattern, a widely accepted idea is explained and then a specific, usually new, application is linked to it. Basically, this technique allows you to take the listener from what is generally known (and accepted) to some unique (and not well-accepted) conclusion. A speech so organized might work in gaining acceptance for new products such as thermoelectrics. These modules are small conversion systems that take readily available heat energy and convert it to electrical energy. They have been used widely for microwave repeater stations in remote zones and are recognized as highly reliable, small, and quiet devices. Building upon this general recognition, you might argue for their use in specific applications such as long-distance truck heaters. Their small size, reliability, and quietness could be seen as equally applicable to trucking as it is to remote power applications.

Problem to Solution A widely used pattern for technical communication is the description of a problem and then its advocated solution. You might analyze some difficulty such as the need for early detection of storms, early preparedness for natural disasters, mineral assessment of proposed mining sites, and spotting devices for large-area crop health. Your solution to these seemingly disparate problems might be the proposal for Landsat, a satellite capable of complex imaging of the ground and sky for identification purposes.

Because the *Report Form* is a version of the problem-to-solution organizational pattern, and the report form is so important to the corporate world, we are going to look at it in detail. You are probably already familiar with the written report form, as it is the prevalent pattern for scientific or experimental laboratory work. Because of your familiarity, it might be worthwhile to compare the written report version with the oral version as it generally appears in corporate practice:

Written Report Form

- Purpose and Problem Statement
- Summary
  1. Statement of findings
  2. Recommended action based on findings
  3. Bottom line costs based on plan
  4. Elapsed time for program
- Review of Existent Work and Background

- Methodology
- Results of Experimentation and Testing
- Discussion of Methodology and Results
  1. Special problems
  2. Unexpected results
- Plan of Action
  1. Proposed solution to any difficulties
  2. Specific budget breakdown
  3. Detailed schedule
  4. Resource requirements
     a. Human
     b. Equipment

The summary section comes first in written reports so that a busy reader can determine at the beginning if this particular report provides a possible answer to the problem under consideration. The summary sketches the most essential information about the test or work. If it seems to have promise, the reader can then examine the remainder of the report for detailed information. If not, a minimum of time is lost in reading through the summary.

Oral Report Form

- Purpose and Problem Statement
- Review of the Existent Literature and Background
- Methodology
- Results
- Discussion
  1. Interpretation of the range of results
  2. Advantages/disadvantages of approach
  3. Trends that results seem to indicate
- Recommended Action Plan
  1. Refinements of overall procedures
  2. Recommended changes in design
  3. Budget and schedule
     a. Human resource allocations
     b. Equipment allocations
  4. Program management
- Request for Acceptance of Proposal

Obviously, the same proposal is presented differently in the written and oral channels. As a communication scholar once said, "A speech is not an essay standing on its hind legs and talking." The written report and the oral presentation contain much of the same information, but they are different events. Looking at their organization lets you see just how that difference works. Listeners come to an oral encounter expecting to hear the speaker make the case completely. Starting with a summary would not fit what we know about in-person communication. Can you imagine a group of people stopping a speaker after a summary to say, "Just a minute. We don't want to hear any more of this." Even in the time-conscious world of corporate business, such an action would be rare. A reader's distance from the writer, however, allows the reader to discard the written report after reading a few sentences.

Motivational Pattern One of the most widely reproduced organizational patterns for speeches was proposed by Alan H. Monroe in the mid-1930s and is called the "Motivated Sequence." It draws on the concepts of human attention and need.[4] The popularity of this pattern and its usefulness make it worth looking at in detail. It has five basic steps:

- *Attention:* Creation of interest and desire.
- *Need:* The development of the problem, through analyzing things wrong in the world and relating those wrongs to an individual's interests, wants, or desires.
- *Satisfaction:* The proposal of a plan of action that will alleviate the problem and satisfy the individuals' interests, wants, or desires.
- *Visualization:* The verbal depiction of the world as it will look if the plan is put into operation.
- *Action:* The final call for personal commitments and deeds.[5]

Understanding this pattern will necessitate your reviewing the section on human needs in Chapter 5. The second step requires you to select some need(s) to be raised in your audience. Think of it this way: if you wanted to sell a home-owner fifty square feet of solar panels for domestic hot water, what need(s) might you play upon? With the threat to affordable energy, perhaps the need for security would be appropriate, and with some receivers, the need for self-esteem would be the right choice. Your pattern might look like this:

Attention

An attention grabber such as:

- Interesting statistic, "Would saving 50–70 percent on your hot water costs be attractive to you?"
- Story or little known fact, "In the 1920s and 1930s solar collectors were common in Florida. So much for 'modern technology.'"
- Quotation, "My friend from Israel said, 'Do you Americans have so much energy that you can afford to waste the heat of the sun? My country uses every bit; solar collectors are everywhere. Where are yours?'"

Need

Detail the problem as relevant to the listener's fears (security) and self-esteem. Analysis should include:

- What is the energy-shortage problem and how does it harm the individual's security and self-esteem?
- What has caused the problem and what has this to do with the listener?
- Will the problem grow?
- How can the harms to the listener be documented?

Satisfaction

The plan is detailed here. It should include:

- What kind of equipment is needed?
- How will it be installed?
- How much will it cost?
- Will it need to change the look of the house?
- How long does installation take?

Visualization

Help the listener to picture the equipment and its daily working.

- Here's how the collectors work.
- Here's what you do (or don't do).
- Here's what they'll look like.
- Here's what the neighbors will think.

Action

Tell the listeners specifically what you want them to do.

- "Sign this contract with Solar, Inc. for the purchase and installation of three solar panels and supporting equipment."

Monroe's Motivated Sequence is extremely versatile. It guides a speaker toward placing materials where they might do the most good. For certain topics and audiences, you might find it a most useful organization.

Patterns, Purposes, and Audiences

We have been discussing preparation for formal speaking situations in this chapter because they are common events for technical professionals. Much of what we have been saying is also true of other oral encounters in daily life.

In a very real sense, both informal and social communication involve preparation, thinking about goals, and determining specific means of organization. These steps are completed so casually that most of us do not even recognize them. Think about the last party you attended. Did you wonder who was going to be there? Did you want the others to like and accept you? Did you worry about what you might say? If so, you were preparing for the encounter. That preparation is so common that most of us overlook it.

How about research? Surely nobody researches before stopping for coffee with friends or meeting a colleague for lunch. Most of us do try, however, to keep up with topics interesting to those we value. If you know nothing about music and are dating a music enthusiast, you will probably find yourself digging up information on Van Halen or Beethoven. People who have nothing to say to each other usually do not stay together for long. As a result, many of us search out topics that interest those who interest us.

WRAP-UP

There are no formulas that will tell you how to decide on a topic and construct a set of words and gestures to help you present it. Neither are there any quick ways to researching information that will give your speech substance. There are, however, some guidelines one can follow to select, research, and organize a topic. We have suggested a number of places to look for information and six organizational patterns for arranging it.

The one commonplace that seems to be true of all speech preparation is that the audience determines whether the pattern was good, the information acceptable, and the goal reached. Your splendid and highly successful speech given to a different audience at some other time might be a dismal failure.

Sometimes people look for easy paths to oral competence and social success. They want a "speech for all occasions," "eloquence in three lessons," or the "sure-fire phrase to attract the best-looking person in the room." Unfortunately, such a market encourages those who make false promises. Unless a training program or piece of advice shows you how to work through entropy potentials and recognizes the hard work oral communication involves, save your money. There really are no shortcuts.

ACTIVITIES

1. Compile a list of professional organizations and their journals for the technical discipline in which you are majoring. Then:
    a. Check your campus library to see if the journals are in-house.
    b. Ask your major-department chairperson to request purchase for your library of any that are missing and seem important to your professional learning.
    c. Examine the journals for notification of yearly conventions, regional meetings, or special services.
    d. Join at least one of the organizations as a student member.
2. Ask your speech communication professor to arrange with your campus library a special tour/workshop for the purpose of understanding underused library services such as interlibrary loans, government documents, computer searches, and little-known resource materials.
3. Arrange an interview with a corporate professional in your discipline on a topic you are researching for classroom presentation. Prepare for that interview carefully and use the information you gain in your intended presentation.
4. Select and gather materials on a topic of some controversy within your discipline. Topics might be something like:

    - Nuclear Power Can Be Used for Peaceful Purposes

- Animal Testing and Experimentation Within Universities and Industries Is Necessary (Is Not Necessary)
- Safeguarding Privacy in Light of Growing Computer-Stored Data

Determine your immediate and eventual goals. Organize the body of your speech according to the pattern that best helps you reach those goals. Try:

- chronological organization
- spatial organization
- pro-con development organization
- general-to-specific organization
- problem-to-solution organization
- motivational-pattern organization

5. Listen to one of the lecturers in any of your technical courses. Analyze the lecture by speculating on:

   - The instructor's immediate and eventual goals for you.
   - The kind of research needed and actually used to prepare the lecture.
   - The kind of organizational pattern used.
   - The kind of organizational pattern that should have been used if you feel the lecture was not particularly successful.

NOTES

1. Frederick Williams, *The New Communications* (Belmont, Calif.: Wadsworth, 1984). See Chapter 15.
2. For descriptions of these two orators, see Mildred Freburg Berry, "Abraham Lincoln: His Development in the Skills of the Platform," in *A History and Criticism of American Public Address*, vol. II, ed. William Norwood Brigance (New York: Russell and Russell, 1943), 828–58, and Myron G. Phillips, "William Jennings Bryan," in *A History and Criticism of American Public Address*, vol. II, ed. William Norwood Brigance (New York: Russell and Russell, 1943), 891–918.
3. For a summary of these studies, see Arthur R. Cohen, *Attitude Change and Social Influence* (New York: Basic Books, 1964), 2–6.
4. For the most recent version of the sequence and a discussion of its use, see Douglas Ehninger, Bruce E. Gronbeck, and Alan H. Monroe, *Principles of Speech Communication*, 9th brief edition (Glenview, Ill.: Scott, Foresman, 1984), 254–62.
5. Ehninger, Gronbeck, and Monroe, *Principles*, 249.

CHAPTER EIGHT

PROVING: EVIDENCE AND REASONING

In each of the following three stories, notice how the sender assumes the burden of proving the assertions being made:

Edward Jenner, "The Theory of Vaccination," 1798:

Although I presume it may be unnecessary to produce testimony in support of my assertion "that Cow-pox protects the human constitution from the infection of the Small-pox," yet it affords me considerable satisfaction to say that Lord Somerville, the president of the Board of Agriculture, to whom this paper was shown by Sir Joseph Banks, has found upon inquiry that statements were confirmed by the concurring testimony of Mr. Dolland, a surgeon, who resides in a dairy country remote from this, in which these observations were made.[1]

Interview by H.J.W. Dam with the (then) unknown Wilhelm Conrad Roentgen, regarding the "mysterious" cathode rays, 1893:

"Now Professor," said I, "will you tell me the history of the discovery?"

"There is no history," he said. "I have been for a long time interested in the problem of the cathode rays from a vacuum tube as studied by Hertz and Lenard. I had followed theirs and other researches with great interest, and determined as soon as I had the time, to make some researches of my own. This time I found at the close of last October. I had been at work for some days when I discovered something new."

"What was the date?"

"The eighth of November."

"And what was the discovery?"

"I was working with a Crookes tube covered by a shield of black cardboard. A piece of barium platinocyanide paper lay on the bench there. I had been passing a current through the tube, and I noticed a peculiar black line across the paper."

"What of that?"

"The effect was one which could only be produced in ordinary parlance, by the passage of light. No light could come from the tube because the shield which covered it was impervious to any light known, even that of the electric arc."

"And what did you think?"

"I did not think: I investigated. I assumed that the effect must have come from the tube, since its characteristics indicated that it could come from nowhere else. I tested it. In a few minutes there was no doubt about it. Rays were coming from the tube which had a luminescent effect upon the paper. I tried it successfully at greater distances, even at two meters. It seemed at first a new kind of invisible light. It was clearly something new, something unrecorded."[2]

You and a Friend in casual conversation, yesterday:

Friend: "Don't waste your time on an M.S. in geology. There's no future in it."

You: "What do you mean? I've had a 3.6 GPA since sophomore year. I'll get into grad school easy, and I'm set."

Friend: "I'm not talking about what kind of brain you are. Take your 3.6 and apply to a good M.B.A. program. It's a smart move."

You: "Who says?"

Friend: "Look at the latest Fortune Magazine. *Do you know what it says M.B.A. graduates make as a beginning salary?"*

You: "So tell me."

Friend: "Plenty. My friend, Lisa, got her M.B.A. and started at BIG Corporation at 35K. You remember Terry? He's that hotshot geologist. Graduated last year. Still has no job."

You: "Hmmm."

All three stories deal with the acceptance of ideas, and proof is at issue. Persistent in human communication is the challenge to answer the demands "Show me," "Who says so?" and "How did you find it?" Jenner calls on credible experts of his day to back what he thinks he has found; Roentgen describes his procedure and conclusions step-by-step; Friend builds a case based on *Fortune* and the experiences of Lisa and Terry. All are asking for acceptance of some conclusion, and all are using proof to gain that acceptance.

The concept of proving is inherent in communication; but, in many ways, it is popularly misunderstood. Stereotypes of someone reciting a data table or citing a learned expert come to mind as proof.

They might be, but the most important component is left out of the description. Proof is decided in the mind of the receiver. A *listener* studies the data table and finds it convincing or hears the words of the expert and rejects them. The decision as to what is or is not proof requires a judgmental process. Erwin P. Bettinghaus says that "When listeners or readers believe, the communicator has proved his case."[3]

Suppose you listened to Friend, considered the *Fortune* article and the Lisa/Terry story, and then concluded, "So what? Money has nothing to do with what I want from a degree." Suppose the medical profession of Jenner's day had decided his experiments and the supporting statements of Lord Somerville were unconvincing? Suppose Roentgen's critics had found fault with the way he demonstrated his cathode ray theory? None of the three senders would have proved the case.

Seeing proof as a decision made by the receiver rather than as a straight recitation of numbers, cases, or quoted statements forces a communicator to consider audience response. In this chapter, we plan to explore the components of proof, its sources, its proper uses, and those cases that are counterfeit.

COMPONENTS OF PROOF

The British mathematician Jacob Bronowski says, "In order to act in a scientific manner, in order to act in a human manner at all, two things are necessary: fact and thought."[4] These two components, these activities that qualify us as actors in a human manner, are similar to the traditional way people have thought proof works.

Proof consists of evidence and reasoning. Used together, with care taken that each is legitimate and appropriate to the receiver, evidence and reasoning allow a communicator to increase the probability that a message will be acceptable. Missing evidence ("Why should I show you? It just makes sense.") or missing reasoning ("What have those numbers to do with what we're talking about?") permits a great deal of entropy. The basis for belief seems weak; the listener becomes resistant; the message is judged questionable.

The scientific community is bound by demands for proof. The scientific method itself is both empirical and tied to rational conclusions. In describing the scientific method in his book, *Reasoning With Statistics*, Frederick Williams notes:

For one thing, this approach [scientific method] *places stress on gaining knowledge through the process of observation. Whatever is said about behavior is reasoned from systematic observation and is tested and retested by observation.*[5]

No person pretending to be a technical professional in the scientific community can afford to misunderstand how evidence and reasoning function.

SOURCES OF PROOF

As you can imagine, there are many theories of how proofs work. Aristotle's speculations on proof are still thought to be sound, and many modern theories are rooted in his conclusions. As we noted in Chapter 6, Aristotle postulated that all communication, which he called "rhetoric," is persuasion. Since he equated rhetoric with the goal of various audience responses, his translator Lane Cooper, in turn, equates persuasion with proof.[6] Just as we have argued that proof involves a decision on the part of the receiver, so Cooper sees Aristotle using the terms *proof* and *persuasion* synonymously.

Proofs (persuasions), Aristotle said, are of two kinds: artistic and nonartistic. Those that are nonartistic are found in witnesses, contracts, and admissions secured under torture.[7] The communicator does nothing to secure these. They are simply "out there" to be used. Artistic proofs, on the other hand, are created by the communicator through the speech itself. There are three of these: ethos, pathos, and logos.

Ethos

We have discussed *ethos* under its modern name, credibility. We urge you to review that material in Chapter 6, for a person's ethos is really an interactive process. A person *does* something, and the receiver decides if it is credible or not. Aristotle thought that the character of the speaker is one vital element in securing the listener's response.

Pathos

Pathos is the kind of proof elicited through a state of emotion inspired by the speech. A person's reactions to anything vary depending on the state of feeling, "for we give very different decisions under the sway of pain or joy, and liking or hatred."[8] A speaker who makes us weep, laugh, or feel an emotional response of any kind is using pathos to secure our acceptance. Maslow's Hierarchy of Needs model is the modern counterpart to Aristotle's recognition of pathos.

Logos

Aristotle called the thinking and reasoning aspect of proof *logos*. People want and, in fact, demand solid content and legitimate argument. When speakers use logos, Aristotle believed they make it possible to "demonstrate the truth."[9]

Putting these sources together, we see proof as the evidence (largely nonartistic occurrences the speaker finds ready-made) chosen and used for pathos (artistic proof to make the audience *feel*) and/or logos (artistic proof to make the audience *think*). The chosen pieces of evidence are linked in the reasoning process and structured through the presentation of the communicator (largely artistic) whose personal character (ethos) is judged by the listener and helps or hinders the decision made.

EVIDENCE: A DOCUMENTATION FUNCTION

Evidence is the verifiable and empirical part of proving. It involves the raw material of reports: that someone said something or followed some plan of action. It asserts that some event actually did happen or actually does exist in the manner described. An outsider must be able to check the accuracy of the report or the statement. Evidence is usually divided into facts and opinions.

Facts

Please notice that the use of the word *fact* here differs from the use described in Chapter 3. There, quoting the meaning given to this term by the General Semanticists, we said that one way of defining fact statements is to consider them possible only for the person who directly observed the event. Here, in a looser definition of fact used by those writing about evidence, we consider that *any* person's direct observation can be classified as fact. For your purposes now, consider that anything you call factual must have been perceived directly by someone. In this way, every experiment, every description of methodology, every statistical classification, is assumed to have happened. Someone observed it, performed it, went through the steps, or counted the individuals later classified in the statistical summary. These are the activities called fact.

Remember that, when you present factual proof, you are asserting that the event actually happened and is being described accurately. Examples of fact evidence might be:

- Experiments run to see what will happen
- Tests conducted to observe the limits of the object
- Analyses made of certain substances
- Persons or objects counted to determine a range
- Persons or objects presented as part of some other event
- Numbers or mathematical symbols used according to established methodology

In each of these cases, the existence and accuracy of the event in question is being assumed. Each event or number is capable of verification by another person. The experiment can be run and the test conducted again. Reports of the original procedures can be examined and duplicated in analyses. The same persons/objects might be counted again or similar ones counted for the first time to check reliability. Whether the person (a witness to a crime, for example) or object (the gun used to commit that crime) actually was involved in the event in question has to be demonstrated. Any numeric procedure, whether it is the summarizing of the counted persons/objects into statistical categories or the use of a complex mathematical formula, must be established as part of an accepted methodology.

These pieces of evidence are a necessary part of daily communication, since you can never forecast when someone might challenge, "How did you come up with that idea?" As technical professionals working in the corporate world, your command of such evidence is expected. Look again at Chapter 7. Notice how the body of any presentation presupposes the ability to present factual evidence. Each report form assumes that a communicator can document what is said and that the documentation actually exists.

Opinions

Evidence can also be opinion. An opinion, a judgment made by someone concerning something, describes how a person thinks and feels about some event. Factual evidence often does not exist, and communicators need to offer instead the verifiable statements that represent someone else's viewpoints. When you present such judgments, you are asserting that the person cited actually does hold such views and did make some statements regarding them. In this sense, opinions are both verifiable and accurate.

Acceptance of opinion evidence rests upon the credibility the receiver gives the person who makes the judgment. Receivers tend to suspend credibility for those who are assumed not to know. Your music-major friend may be a bright and honest person, but you would probably be suspicious of anything that friend says about subatomic particle structure. The marketing specialist in your corporation might respect your opinions regarding technological subjects, but would doubt your conclusions about the right way to run a persuasive campaign so that your product sells to the general public. We regard the acceptance of a person's opinions based on individual expertise as extremely important. Chapter 14 is devoted to helping you detect and prepare to be acceptable experts.

Strong and Weak Documentation

Neither the academic nor the corporate world tolerates weak evidence. Nothing can mark a young technologist as unprofessional more quickly than the reputation for presenting flawed evidence. Entire contracts sometimes rest on the customer's acceptance of test or experimental results. If these are proved to be faulty, whether through error or carelessness, careers are jeopardized. Forgery or distortion of data amounts to an invitation to professional suicide.

Presentation of the strongest possible factual and opinion evidence is the goal of corporate research. It should be no different for the student during the academic years. Whenever you consider any evidence, fact or opinion, as documentation for some conclusion, make certain it passes the following six tests.

1. Is the Evidence Reported Accurately? The sender of a message is responsible for being certain of its accuracy. Every quoted word or number must be copied carefully. Whether you are working from your own experimental data or those of another, it is easy to make an error in reading or recording.

Whatever is said or measured originally must be stated without distortion. Ethical and pragmatic considerations should convince you that half-truths or quotations taken out of context have no place in research. Fabrication is a matter of such serious consequence that we dismiss it here as beyond contempt.

2. Is the Evidence Consistent Internally and Externally? Beware of possible internal consistency problems. Whether it is a test run that seems to come out differently every time it is tried or expert opinion that varies from beginning to middle to end of the same report, something is wrong. If differences in the test cannot be explained or shifts in the opinions of your expert unraveled, do not use them. To present a single test result from a variable series or a single portion of the opinion, as if no differing conclusions exist, is foolish at best and distorting at worst.

Likewise, be cautious if the evidence is not consistent with other external sources. Many of the examples we have used throughout these chapters speak of discoveries that were different from other work of the same time period. None of these discoveries, however, happened in isolation. It is rare, if at all possible, for someone to come up with an absolutely novel finding that nobody has thought about before. Most discoveries are the story of several people close to finding the same answer. One person happens to compile the needed proof first and the discovery is attributed to that one. Be on your guard when your conclusions or those of someone you quote are vastly at odds with what everybody else is saying. You may have identified a true giant of science, but most likely you have a finding that is somehow flawed.

3. Is the Evidence Directly Relevant to the Point Under Contention? Make certain that the fact or opinion used is pertinent to your discussion. An author might be using the same words you use and yet mean something different by them. Remember that verbal symbols operate at different levels of abstraction. Highly abstract words have multiple meanings and are always prone to confusion. Consider the following direct quotation:

The stronger of two men, we say, is the one who can lift the heavier stone or stretch the stiffer bow. This measure of force, with which Ulysses established his right among the suitors, and which indeed plays a great part in the stories of ancient heroes, already contains the germ of the objectivization of the subjective feeling of exertion.[10]

Is the author a literary critic? He speaks of "subjective feelings," "exertion," and refers to classic myths. These are the words critics use in reference to literature, yet they come from the pen of Max Born explaining the theory of relativity. A hasty conclusion that some author is using terms the same way you are might mislead you in any number of quotations. Be careful.

4. Is the Methodology Used to Get the Original Conclusions Sound and Accepted? We spoke in Chapter 6 about the need for acceptable methodology. Your evidence might and probably should be rejected if the procedures used to gather the data are suspect. Perhaps the statistical manipulation was illegitimate, or the reporting conditions were irregular, or some necessary checks and balances were left out of the experiment. Every discipline has sets of acceptable ways its researchers must follow to demonstrate their conclusions. Evidence in violation of these ways is suspect. Make certain that your own data were gathered according to the best conditions possible, and check to see that anyone you quote was equally careful. A report that does not detail the method used to gather evidence and form conclusions is a questionable piece of work. Avoid such researchers, and find your evidence among those who follow strict guidelines.

Be extremely careful of mathematical manipulations that might be used incorrectly. Statistical analysis requires detailed steps before conclusions are reached. Studying the sample, its representative members, and how the sample was chosen are just as important as the numeric handling of these variables. So it is with other mathematical processes. Examine the techniques carefully. If anything looks suspicious about any part of the process, choose another study.

5. Is the Source of the Evidence Credible? Numbers published by a person who has a reputation for fabricating data will probably not help you build your case. Neither will statements made by an expert who lies or whose comments change depending on what the highest bidder wants. Publications known for shoddy work or careless handling of details should not be used for your research. Remember that your receiver has to judge your evidence as worthy of belief. Many fine publications and highly capable individuals exist. Search them out for your evidence.

6. Is the Evidence Recent? The modern world and especially the scientific community within it change rapidly. It is extremely embarrassing to cite some piece of information only to find that it has been denied or altered in a later publication. One reason that the report form, both oral and written, includes a "review of the literature/

review of the work" step is to force recognition of what others have done. Listeners and readers expect to know when you stopped your survey of what had been happening.

In most disciplines, you need to find the most recent evidence. Techniques change, laws are altered, countries shift their allegiances, new leaders take office, and persons quoting last year's materials are sadly out of touch. If nothing has been done with some concept since the time you cite, or if you quote a landmark thinker for historic perspective, old references are legitimate. Be cautious when you suspect that new thinking has superseded the old.

If your facts or opinions can withstand the tests suggested, they are probably strong documentation for your point. With each test that the evidence fails, its strength diminishes, and your ability to persuade your receiver decreases. Using weak evidence is a dangerous waste of time. It is unlikely to be seen as proof, and it leaves your listener wondering about your competence as a researcher.

REASONING: A LINKING FUNCTION

Evidence is only one half of the process of proving. Consider what happens in your thinking process when you hear a crash in the next room and, upon investigation, see your coffee mug smashed on the ground and your cat cowering in the corner. "Bad cat," you reason. "The mug was on the shelf where she likes to sleep. She must have knocked it down." There is no overt connection between a mug on the shelf, a cat who likes to sleep there, and the same broken mug—except that your mind links these separate events together. You know, because of past observations, that the first two events happened. Your mind puts them together and draws a conclusion that you can substantiate in no other way. If someone challenged you, your response would be something like, "But it makes sense. Mugs don't jump down from shelves by themselves. Something must have knocked it down. I've told that cat a million times to stay off that shelf. She goes there anyway and then always hides in the corner. She did it all right."

You have put together several separate ideas in order to draw a conclusion. This process of linking known information in order to speculate on the unknown is called *reasoning*. Reasoning "hooks up" bits of evidence, sometimes in strong fashion, sometimes in weak fashion, and forms information, ideas, or concepts different from those with which you started.

Arthur N. Kruger, in his book on debate, points out that the French astronomer Leverrier based his proof of the existence of an undetected planet on reasoning.[11] He linked together (reasoned) known facts about Newton's law of gravitation and deviations of the planet Uranus from its observed orbit, and he concluded that an undiscovered planet was causing the deviation. He had no direct evidence, for the machinery that later detected the planet Neptune had not yet been developed. His proof was a cause-and-effect reasoning process.

Linking ideas to other ideas in order to draw novel conclusions is a basic human activity. The discipline of logic has been studied for centuries as a way of learning to think "correctly." The study is worthwhile, but you must keep in mind that it is the receiver of your message who determines whether the thinking is straight or skewed. All of your protest that, "This is a valid conclusion," comes to nothing if the receiver does not agree. You can easily imagine a historic scene where a prisoner objects, "But, Your Highness, that conclusion cannot be drawn logically from these premises," and the king responds, "Who says it can't? Off with your head."

No one can guarantee that rational reasoning will win acceptance of your proof, but the absence of good reasoning definitely places that acceptance in continual question. Professional disciplines are filled with practitioners who test and retest the proof offered by one of their members. If the portion of proof that is based on reasoning appears weak, the conclusion has little chance of gaining support. In this next section we present five specific types of reasoning that are of special use to the technical world; and we note those forms of fallacious reasoning you must guard against.

Strong Links: Categories and Types of Reasoning

When speaking of reasoning, people traditionally refer to the concept of argument. That makes sense, for while reasoning is the formation of a conclusion through the linkage of ideas, an argument, in much the same way, leads a listener to some final summary through a series of prior assumptions. Traditionally, argument and reasoning have also been divided into two general umbrella *categories:* inductive and deductive. These divisions have been the subject of controversy for thousands of years. Some have argued that one is better than the other, that one is the more reliable, that only one can bring you to "truth." Our position is that such conclusions are largely pointless.

Both categories of reasoning are used widely; both bring possibilities of information that the other does not; both need study, since neither can be avoided.

Inductive reasoning requires observation of specific events before any general conclusion about the whole class of events is possible. It is the "from specific to general" pattern. If you studied a diamond and noted that it cuts glass, you might conclude that *this* particular diamond cuts glass. If you continued your study, noting that each diamond you examine cuts glass too, you might be led to assume that *every* diamond cuts glass. Based on your evaluation of specific members of a class of substances (diamond 1, diamond 2, and so on), you conclude something about the entire class (all diamonds). Science, as an empiric discipline, is based on observation. It is at heart a system of inductive event processing.

Deductive reasoning assumes that general knowledge about a whole class of things makes it unnecessary to examine any single member of that class. This is usually called the "general to specific" pattern. If you know that all dogs bark, for example, it is a waste of time to check the barking capacity of a given dog. You can assume without examination that, if it is a dog, it can bark.

Scientists use deductive reasoning widely despite their reliance on observation. You do not need to examine the force of gravity on a given piece of metal each time it is used in a test, for example. Knowing what the gravitational pull is on all metal objects, you can conclude what it is for any *particular* one, without actually measuring it.

There are many specific *types* of reasoning. If you were a debater, logician, or prospective lawyer, you would benefit from studying all these different varieties. They are beyond the scope of this book. We have selected five of the most common types, assuming that these are the ones technical professionals, as well as most people, use frequently. Their correct use will help decrease the probability of receiver entropy.

1. Cause and Effect Causal linkage connects two or more things in a special way: the one (or more) is the direct reason why the other(s) happened. The question here is, "What *produced* this phenomenon?" Sometimes, perhaps most frequently in the scientific community, the process of causal reasoning works from effect to cause. Some event is observed—a planet deviates from its assumed position, a volcano erupts, a child dies of a spotted disease—and the observer wonders why such things happen. Consider the answers possible for any of these events:

- The proper ritual was not observed by the community.
- An unseen planet is pulling the observed one.
- A person walked on the sacred mountain without permission, angering the fire god.
- The buildup of underground pressures found a fissure in the earth's crust.
- A witch gave the child the evil eye.
- Tiny microbes invaded the body, killing it.

Whether the question is from cause to effect or effect to cause, the history of science has been the probing of what something does or why something happens.

2. Sign to Conclusion A sign is an observable clue, whose presence makes it possible to reach a conclusion. Seeing the lights on at your friend's house and the car parked in the driveway are clues that your friend is at home. Basic in the reasoning from sign is the assumption that certain things usually happen along with other things. Notice that no claim is made that the one *causes* the other: neither the lights being on nor the car parked in the driveway caused your friend to be at home. Past experience tells you that, when house lights and cars are observed, it is reasonable to assume habitation.

Signs are extremely common bases from which we draw conclusions. Austin Freeley notes that care must be taken to differentiate between *reciprocal* and *nonreciprocal* signs. The first occurs when the two variables work both ways: each is a sign of the other. They are nonreciprocal when one variable is a reliable sign of the other, but the second is not a sign of the first. His example of a nonreciprocal sign is that, if a man is President of the United States, we may take this as a sign that he is at least thirty-five years of age. We cannot assume that because a man is thirty-five, he is president.[12] Our friend-is-at-home example is nonreciprocal. The lights on and car in the driveway entitle us to think that the friend is present. The friend, however, might have parked the car in the garage and is sitting in the dark too.

3. Example Examples are assertions that some event has happened at least one or more times and that it is therefore legitimate to link these cases to the assumption that the event will happen again. You usually need a number of cases to establish that a trend is present. If a given experimental technique has been documented in a certain number of cases as being dangerous because of incidences of fire, you are entitled to conclude that use of that technique in your experiment increases the possibility of fire hazard. You would probably be unimpressed if only one case of fire resulting from use of the technique were reported.

What would happen, however, if a single case of some startling phenomenon came to light? The landing of a single spaceship on the moon was enough to warrant the conclusion that space travel and landings are possible human activities. The number of examples it takes to gain acceptance is not some magic digit. In some instances, a single case is sufficient; in others, the sender has to decide how many cases it will take to increase the likelihood of acceptance.

4. Analogy An analogy is a comparison between two similar things. Because something is known about object A, you may reasonably assume that the same something is also true of object B, which is similar. This mode of thinking is based on the assumption that similarities are widespread. If some common elements can be demonstrated between objects, it seems right to conclude that others are also possible.

In many ways, analogy functions along the "It happened to him, so it might happen to me too," pattern. If you and a coworker were hired at about the same time, received approximately the same evaluations of your work, are the same age, and have participated in the same number of equal-status projects, you have many documented similarities. When the coworker receives a nice promotion, you would probably expect to be promoted soon also. Your conclusion is drawn from analogy.

5. Authority In reasoning from authority, you assume that a conclusion is valid because the "Special One" says it is. It takes intense personal audience analysis to discover whom a listener thinks is authoritative, but most of us are susceptible to such thinking. At one time in our lives, it was enough to know that, "Mommy says so." Later, the same reasoning became, "Professor Berriman told me."

Acceptance of an authority derives from many sources. The authority might be the perceived power source or have specialized knowledge or be associated with a particular mystique. For whatever reason, the authority has such status with the receiver that conclusions will be accepted merely because the authority is their source.

It is not necessary for the authority to be a single person; indeed, it need not be a person at all. Some of us approve or disapprove ideas because, "Nature supports such a view," or "Science can never accept such a conclusion." The credibility that constructs such as nature, science, religion, education, etc., have for some of us is enough to verify ideas. Many persons stop the critical process when told that, "The Bible says it," or "The Republican Party believes that this is the right way," or "The University supports this program." In each of these cases, the listener persuaded by such reasoning is bowing to the weight of authority. The conclusion is acceptable because the authority supports it.

Weak Links: Fallacies

Try to identify correctly the type of reasoning at work in the following five examples. Can you also sense what is wrong with each one?

1. The people in the apartment next door have a place just like ours. They have one child, just as we do. Last fall they put weather stripping around their windows, and their heating bill went down nearly 10 percent. We ought to weather strip. I'd like smaller fuel bills too. (Analogy)
2. There's a lot of heart disease in my family. I'm not going to eat salt anymore, and I'll eliminate the chance of a heart attack. (Cause and effect)
3. The president of our fraternity says we should boycott graduation in protest of the administration's ban on our fund-raiser. It's a shame, but I'm not going to my own graduation. (Authority)
4. Lynn didn't study for the math test and she got an A anyway. I heard Tyrone say that neither he nor Gene studied either, and I know both of them got A's too. Next time I'm not going to study at all. I need an A. (Example)
5. My throat hurts, and my eyes are watering. I think I'm getting a cold. (Sign)

If you classified these types correctly, you have been paying attention. If you thought some of these statements seemed right, you have a lot of company. Each commits an error known as *fallacy*, and the hallmark of fallacies is that they masquerade as legitimate reasonings.

Fallacies, as Fernside and Holther say, are the counterfeits of argument.[13] In one sense, their widespread use can be criticized because they make arguments "invalid" or "permit illegitimate conclusions." These are serious charges, but there is yet another that we find even more disturbing.

In his book on logical thought, Richard Purtill says that a fallacy, "is an argument that in fact is not a good argument, but for some reason appears convincing."[14] This appearance of being convincing should put a communicator on guard. We have asserted throughout this book that the aim of good communication is to keep entropy at the lowest level possible so that an accurate message has a chance of getting through. Fallacies are purposeful distortions in the message. It is as if a short circuit were knowingly designed into the system. The listener, unaware of the flaw, finds the message convincing without having the opportunity to judge it on its actual basis.

Because fallacies have been propounded—and resisted—throughout the centuries, much can be said about them. A thorough study would mean learning nearly a hundred fallacies and their variations. Many of them have been given Latin names, such as *post hoc, ergo propter hoc* or *ignoratio elenchi*. These names might impress your friends, but our purpose is to make you recognize the central question about fallacies: just what is it that happens to you when a sender purposefully uses fallacy to increase listener entropy?

We have selected ten of the most common fallacies, called them by their English names, and explained each briefly from the sender's point of view. Each is also accompanied by an example to help you identify it when next you meet it.

- *It Does Not Follow.* My conclusion really has nothing to do with anything else I have been saying. I'll say it anyway.
  Example: Stop your studying and come to the movie with us. Professor Elliott is a jerk for giving a midsemester exam tomorrow.
- *Hasty Generalization.* I really don't have enough data to draw the conclusion I'm making. Maybe nobody will notice.
  Example: Those folks in the metallurgy section are really a bunch of snobs. Just watch Brian walking around with

his nose in the air, not talking to anybody. You'll see what I mean.

- *Circular Reasoning.* I haven't proved my conclusion yet, but I'll just pretend I did and see if anyone objects.
  Example: Of course Myra's smart. Just look at her; she looks smart.
- *After This, Therefore Because Of This.* First this one thing happened, then this second thing happened. It seems obvious to me that the first caused the second.
  Example: I wore my red sweater and won the tennis match. That sweater is my lucky tennis charm.
- *Irrelevant Issue.* This conclusion has nothing to do with what we're talking about, but I'll use this chance to bring the idea up anyway.
  Example: The trouble with you is that you spend too much money. Another thing, I really don't appreciate how you leave me alone so much when we go to a party with your friends.
- *Exaggeration.* I'll pad the data a bit. It makes it all sound a lot more important.
  Example: I just learned that two friends are thinking of divorcing their respective spouses. Everyone I know is getting divorced. Doesn't anybody stay married anymore?
- *Name-calling.* I don't think this argument is going very well. Watch me distract him from the issue by calling him something he really hates.
  Example: I can see how a wimp like you would feel that way about the company's new regulation.
- *Personal Attack.* I can't answer her objection. If I can shift her attention away from what we're talking about with a really sexist remark, maybe she will take the bait and defend herself instead of dealing with the argument.
  Example: You women are all alike. Every time you run into trouble, you bring up this equality stuff. You're a prime example of someone who should stay at home and keep her mouth shut.
- *Word Smokescreen.* I don't want to talk about that. Maybe I can throw up a camouflage of words, and nobody will know the difference.

Example: I'm really glad you asked me about my farm policy. Some of my fondest boyhood memories are of getting up before dawn to help my Dad milk the cows. Those were good and honorable days. People weren't afraid to get their hands dirty with the dignity of hard work. We . . .

- *Appeal to Popular Prejudice.* They seem to like my opponent's idea. I'd better hook it up to something I know they disapprove of. Maybe then the idea will seem less appealing.
  Example: This great plan my opponent is proposing is sadly old and out-of-date. It could have collected retirement benefits when our grandparents were children. I haven't heard one new idea from her yet.

Unless you are on your guard, some of these examples seem to make sense. Their persuasiveness is directed at making you think something has been proved when, in effect, the sender is distracting you from the issue under discussion. Studying "strong links" should give you the knowledge to understand and guard against weak ones. Fallacies abound. We hear them from some of our leaders, read them from billboards, and watch products being sold under their disguise. The way to detect them is to test for their presence. If, despite the fallacy, you decide to vote for the candidate or buy the shampoo, that is your right. It is also your right to know that you are doing it based on a short-circuited reasoning process.

Now that we have looked at some of the common fallacies, consider how you tell a strong reasoning link from one that is counterfeit.

Testing for Strength of Linkage

Fortunately for us, while theorists were inventing Latin names for fallacies and logicians were analyzing reasoning, they evolved some checks and balances. Each type of reasoning is associated with a sophisticated set of tests to see if the linkage is legitimate or counterfeit. Most of these tests are beyond the scope of this book. We have selected a few easily applied questions you might ask yourself when you suspect any of the five types of reasoning. In addition, we direct your attention again to the five examples of fallacious reasoning with which we introduced this section on page 175. When you apply the questions to those five examples, you will probably discover what seemed wrong with them.

Testing Cause and Effect Reasoning

- Is there a direct relationship between the cause and the effect, or is it accidental?
- Does this relationship exist every time the variables interact?
- Are there other causes operative also?
- Is the cause capable of producing this effect?
- Are there solid explanations for any exceptions to this assumed relationship?

If each question can be answered affirmatively, you probably have a linkage between two different events related in a strong causal manner. To the degree that any of the statements must be answered negatively, the conclusion has to be suspect. Our example of the person who is fearful of heart disease eliminating salt from the diet cannot stand the test. The relationship between eating salt and heart disease has been experimentally linked, so the first test is encouraging, but look at what happens to the others. Many other variables contribute to the likelihood of heart attack, yet these were not considered. A person could eliminate all salt, for example, but continue on the way to a heart attack with an inactive, overweight, and pressure-filled life-style. Neither can it be answered affirmatively that the relationship between heart attack and salt exists every time. Some people never limit intake and live long disease-free lives. The example simply cannot hold in its present form.

Testing for Sign Reasoning

- Is the clue inherently related to the event in question?
- Is the interpretation of the clue the only explanation possible?
- Were other variables examined and explained?
- Was an account made of every possible counteracting variable?

Our example of the sore throat and watery eyes cannot withstand an affirmative-answers test. Not all colds inherently involve the symptoms described. Several other explanations might also be legitimate: allergies, smoking, air pollution, or diseased tonsils. The possible counteracting variable of a history of elevated temperature during colds was also not considered. The person *might* have a cold, but the reasoning does not lead to a confident conclusion.

Testing for Example Reasoning

- Is the example widespread and supported by other cases?
- Are the individual members in the example typical?
- Are the individual members in the example diversified?
- Is the example directly related to the point being asserted?
- Can counterexamples be explained?

The conclusion that not studying for a math test is the way to get an A is obviously flawed. The three cases cited are not widespread and neither are they diversified. Notice, that the test of members being *like* the wider population (typicality) is just as important as demonstrating that the members represent diversification. If Lynn, Tyrone, and Gene are very much like others in the math test-taking sample and at the same time represent the variety of persons also in the sample, their cases might be reliable. We could not account for counterexamples in these instances either. How would we explain Pat, Christine, and Mario who studied many hours and did very well on the test? The cases used for the example cannot stand analysis.

Testing for Analogy Reasoning

- Are the things being compared truly similar?
- Are the similarities in crucial-to-the-point areas?
- Have differences in the compared events been acknowledged and explained as uncritical?
- Do other comparisons exist that verify the one being made?

The analogy of comparing energy-cost savings of the people in the apartment next door and those of anyone else must be suspect. It does seem that the similarities of apartment size and number of children make the comparison legitimate. It also seems that the analogy holds for the very thing that energy costs usually consider—that is, size of family and dwelling. The final two tests begin to reveal some areas of questionable content. If we knew that the energy-saving family also lowered its thermostat by four degrees over the entire heating season and made a real effort to keep the door to the unheated hallway closed, while the compared family did neither, we would begin to doubt the 10 percent savings comparison. In addition, if others weather stripped too, but this was the only family in the building to have achieved such a saving, we would see that there must have been other variables at work.

Testing for Authority Reasoning

- Is the authority an expert in the area being questioned?
- Does the expertise come from training, education, and/or unique experience pertinent to the area being questioned?
- Is the authority relatively free from bias?
- Does acceptance of the authority's conclusion rest on any factors of fear or desire for the authority's approval?

Again, the example of authority reasoning rests on doubtful merit. A fraternity president usually has considerable credibility among the organization's members. The source of his credibility may be based on the power he has or the specialized knowledge about the fraternity he is assumed to have. His statements are often accepted as having proved some highly debatable point simply because he said them. When those statements are subjected to the suggested tests, however, they might not stand very well. Whether we could claim expertise for the president is not certain. The young man, we might assume, has no formal training or education in the area of persuasive strategies for demonstrating dissatisfaction. He might claim expertise based on the unique experience of being the fraternity's only president at this particular time, of course. Whether that gives him the basis of analyzing proper persuasive devices such as boycotts is questionable. The last two tests would have to yield a negative response. The young man is obviously biased, as his organization has a lot to gain from liberalizing rules on fund-raising. Finally, there is both the danger of fear and striving for the authority's approval that can mar the legitimate assessment of the president's conclusion. Most organizations frown on members who deviate from accepted practice, and most members wish to retain the approval of their groups. A young fraternity member might feel intimidated enough by his president to accede to the boycott. At the same time, the member might honestly want the president's good regard and that of his brothers. The young man might stay away from his own graduation for reasons that make sense to him, but these are not the links of strong reasoning.

WRAP-UP

Are there ever easy answers? We doubt it, especially when the questions are so complex. Only those who never look beyond the surface of the obvious seem to deal with simple truths. For the educated person and the technical specialist particularly, easy answers should always raise a possible danger signal. The means to probe such dangers involves the demand for analysis of proof. A scientist grows used to establishing good reasons for the community to accept assertions about the world. So should and does the accountant, the public relations expert, or the marketing manager. Together, all these members of a corporate organization work in the peculiarly human activity of proving.

We have described proof as consisting of two parts: evidence and reasoning. Either might stand alone as convincing; but, when both are used together as the best components of a communication, listeners gain the finest compliment possible: that they, unique among other creatures, have the capacity for taking the raw materials of evidence, working them through the system of reasoning linkage, and making a decision as to their strength or weakness. A communicator who can say, "I have proved this point," means that another person has taken a given message, weighed the worthiness, and given it a positive judgment. Weak evidence and fallacious thinking deny the receiver this compliment and testify to the sender's own fears of being able to make a case without deception.

ACTIVITIES

1. Select a fairly long letter to the editor from a magazine or newspaper. What kinds of evidence (if any) does the writer present for the conclusions made? Do you find the evidence convincing? Why?
2. Using the same letter to the editor as you did in Activity 1, describe what kind of reasoning the writer used. Did you find any fallacies? How did you locate them?
3. Identify a position on a controversial subject that you hold strongly. What kind of proof (both evidence and reasoning) would you demand from those holding the opposite view when they spoke about this subject with you?
4. Present a four-to-five minute presentation in class advocating some action. Purposely work in two or three fallacies. When you complete the presentation, ask the class members to identify your fallacies. If they cannot do so, explain what they were and why they count as fallacies.

5. Secure a copy of a U.S. president's speech, either one from the current president or his predecessors. Identify types of evidence and reasoning he used. Test that proof according to the standards established in this chapter. What do your results tell you about that particular president?

NOTES

1. William S. Knickerbocker, ed., "Edward Jenner," in *Classics in Modern Science* (Boston: Beacon Press, 1927), 150.

2. H. J. W. Dam, quoted in Derek J. de Solla Price, *Science Since Babylon* (New Haven: Yale Univ. Press, 1961), 77–78.

3. Erwin P. Bettinghaus, *The Nature of Proof*, 2nd ed. (Indianapolis: Bobbs-Merrill, 1972), 8.

4. J. Bronowski, *The Common Sense of Science* (Cambridge, Mass.: Harvard Univ. Press, 1967), 36.

5. Frederick Williams, *Reasoning With Statistics* (New York: Holt, Rinehart & Winston, 1968), 3.

6. *The Rhetoric of Aristotle*, trans. Lane Cooper (New York: Appleton-Century-Crofts, 1932), 1.2.

7. Aristotle, *Rhetoric*, 1.2.

8. Aristotle, *Rhetoric*, 1.2.

9. Aristotle, *Rhetoric*, 1.2.

10. Max Born, *Einstein's Theory of Relativity* (New York: Dover Publications, 1962), 15.

11. Arthur N. Kruger, *Modern Debate: Its Logic and Strategy* (New York: McGraw-Hill, 1960), 146–47.

12. Austin J. Freeley, *Argumentation and Debate*, 3rd ed. (Belmont, Calif.: Wadsworth, 1971), 127.

13. See W. Ward Fernside and William B. Holther, *Fallacy: The Counterfeit of Argument* (Englewood Cliffs, N.J.: Prentice-Hall, 1959), for a comprehensive treatment of fallacy.

14. Richard Purtill, *Logical Thinking* (New York: Harper & Row, 1972), 55.

CHAPTER NINE

COMMUNICATION AIDS

Communication aids are an integral part of technical and business communication. From the simplest typewriter to the most complex computer-generated chart, visual and audio aids are tools crucial to the technical professional. In this chapter we discuss guidelines for creating effective aids, how to use the best of them in presentations, and the variety of aids available to modern businesses.

AIDS IN TECHNICAL COMMUNICATION

One of our purposes for modeling the communication process (see Figure 2-11) is tracing what happens in the flow of stimuli through channels from sender to receiver and back. This process, in fact, supports a "bigger is better" theory in the sense that, "if one channel is good, two are better." Good communication keeps entropy at its lowest level. A stronger signal reduces the entropy effect better than a weaker signal does. Visual and aural aids increase the strength of the communication signal by providing multiple channels for the stimuli to flow through.

The advantage of multiple channels is demonstrated through a quantitative study measuring the effect of visual aids on audience message retention. Tests showed the following results:[1]

Message Retention

Method	*After 3 Hours*	*After 3 Days*
Oral only	70 percent	10 percent
Visual only	72 percent	20 percent
Oral and visual	85 percent	65 percent

As this study shows, visual aids help people remember messages.

The most common communication aids for businesses are aural and visual, and we will concentrate on these, but other aids exist. The tactile sense helps customers understand a new material as, for example, when people are encouraged to touch and handle a sample. When silicone rubber was introduced for home-repair use, a sample of the material was included on each package to introduce the customer to the unfamiliar material. Food companies use another sense to sell new products. Customer acceptance is based on taste, so prospective buyers get samples and are invited to eat them. Both the silicone rubber and the new food sample are communication aids.

Businesses use communication aids and expect their professionals to be competent in handling them because:

- Aids minimize the distractions present in the environment.
- Aids focus attention on the selected message.
- Aids show a regard for the receiver's response.
- Aids enhance clarity of message.
- Aids increase the efficiency of transmission.

All of these reasons add up to one that is overwhelming: *professionals in business expect communication aids.* Machine aids are a part of daily office routine, and it has become standard practice to use visuals when oral reports are given to peers and management. Even the simplest chart, such as an agenda word chart, is expected at meetings. The higher the level of the people in the organization with whom you communicate, the less time is available. Unless you are the chief executive officer, you will be given a limited and specified slot in a meeting schedule when you will have to influence your superiors. Others will be given a limited opportunity as well. You must make optimum use of the time available to you for telling your story. A variety of visual aids is therefore almost mandatory. Most meeting and conference rooms are furnished as a matter of course with equipment for displaying aids, especially transparencies. Whether you like it or not, your corporate audience will expect well-constructed, competently utilized, and message-supporting communication aids.

GUIDELINES FOR EFFECTIVE AIDS

In an excerpt from his book, *The Elements of Friendly Software Design,* Paul Heckel notes the tenets Walt Disney held regarding "effective communication" in his movies and suggests how they can be used for design of user-friendly computer software.[2] Recognizing that Disney was a master of visuals, we believe some of his guidelines pertain to technical aids as well. When technical professionals speak to an audience, they face the same problem Disney did—designing visual aids that get and hold audience attention. His solutions can help you design materials that work well with audiences. To say that the material content provides its own interest is an inappropriate excuse for poor visuals and shortchanges the total message you provide. We think you can learn a great deal from Disney's guidelines:

Make It Clear

This first tenet, while not mentioned in Heckel's article, is a necessary condition for all message exchanges. Keeping entropy at its minimum is easier said than done. The key is to keep your audience in mind throughout. One of your authors likes to write a single message block per aid and then decides whether the graphic helps or hinders audience comprehension. In this way, he can identify and eliminate issues that support but are not a direct part of the main message. It is best to concentrate on a single idea per graphic aid. Emphasize the main message rather than interject added information and distract the audience from the main event.

Make It Interesting

Clarity does not mean an absence of interesting detail. According to Heckel, Disney used a particular technique to maximize audience interest:

Disney storymen and animators constantly looked for "a piece of business" that would bring a character to life and make it interesting. A character never simply walked from one place to another: too dull. He might be angry and show it in the way he walks; he might drag something along; he might scratch his head. But he would do some piece of business that was in character, advanced the story, and was interesting.[3]

Focusing interest on some graphic aid activates the mind of the receiver and commands attention. A chart with a single letter or word is the least interesting of graphics. Unless the word generates a significant thought process in the mind of the receivers, they will comprehend it too rapidly and will prepare for another idea. A full phrase or sentence provides a modest improvement over a single word, but again the visual will be assimilated rapidly and audience interest will wane. The listener's mind must stay active, with the graphic stimulating clear thought. Color, unexpected print, surprising pictorials, or contrasts can be used to keep the audience tuned to the aid. The aid, in turn, acts as an additional channel of communication, supporting the ideas that are the bottom line of the presentation.

Exaggerate Reality

To heighten interest and help clarity, Disney altered the real world. According to Heckel:

Early animators discovered that two related keys to making a character interesting are exaggeration and caricature. Reality must be the starting point: it is necessary to identify the essence of an animal or a person, and at Disney this was researched in great detail. Once the essence of a character was identified, it was caricatured and its movements were exaggerated. This made the character more interesting than the real one it was based on.[4]

This technique can be applied to visual aids without endangering the decorum of the technical setting. While caricature in a technical presentation is rare and perhaps out of place, examples of exaggeration in visual aids are quite common. Use of bold print emphasizes major messages; magnifying a portion of a sketch can add clarity; scales on graphs may be truncated or expanded to highlight some portion. Each of these techniques exaggerates reality without distracting from the seriousness of the subject matter.

Think in Visual Terms

In another reference to Disney, Heckel quotes Frank Thomas and Ollie Johnson, Disney studio animators, recalling Disney's writing procedure:

Walt usually left out the dialogue until a sequence had been developed to the point where he could see just how little was really needed. If the idea could be communicated with an expression, an action, or a sound effect, or with music, he would not use dialogue. The storyman had to think in visual terms first, and when he did write dialogue it had to tell something about the character and not be merely exposition.[5]

Since we are discussing oral communications where visuals play a supporting role, we do not have the option of no dialogue. We can, however, learn a valuable lesson from Disney's technique. An effective presentation must use visual aids that *support the dialogue but do not contain the dialogue.* Most humans read much faster than they speak. The visual message is observed and absorbed quickly. Few things lose an audience faster than recitation of words on a screen. The audience resorts to using the screen as the source of information and the speaker actually becomes a mere accessory to the communication rather than the source of it.

The visuals should enhance the oral message but never overshadow it. Bold aids are effective when the speaker describes the complete supporting details. Pictures or graphic charts can be excellent when they highlight part of the data, leaving the full interpretation to the speaker. The "think in visual terms" process is an extremely important tool of visual creation, but aids must always play a supporting role to the main attraction.

Prepare the Audience

Audience preparation follows the precept, "Tell the audience what you're going to tell them." We discussed organization and audience attention in Chapter 6. We must remind you that a good sequence of ideas promotes a mind set within the receiver open to the most accurate interpretation of a message.

Many audience preparation devices can be used, and their effectiveness is measured by how well the listeners are prepared to receive the subsequent messages. A very simple and common strategy is to use an agenda chart. The agenda chart can take the form of a broad outline of your presentation. You can use the agenda chart as a "break in ideas announcement" when your presentation moves from one subject to another. Use a pointer or arrows on the chart itself to orient the audience to the next part of the presentation. Too frequently an audience becomes lost when the speaker has changed the subject but the listener is still oriented to the previous idea. Such losses occur because the speaker fails to take the listener along as the presentation moves on. Aids can be used as a visual road map.

Keep Illustrations Uncluttered

Simplicity is a virtue in designing visual aids. Each receiver will attempt to grasp the essence of the visual aid as soon as it is presented. If the visual is simple yet arresting, the audience quickly perceives its essence and prepares to hear the speaker's detailed explanation. If the visual aid is complex, each receiver may concentrate on a different part of the aid and so miss part of the spoken message. A complex visual invites the audience to try to interpret it themselves, thus totally ignoring the speaker.

No aid should ever distract from the speaker. Simplicity keeps this source of entropy minimized. Try going through the message you plan to give with each aid in hand. When you find that there is an uneven distribution of ideas or that you are spending much more of your time on one chart than on others, you should be concerned. Use multiple charts containing a single message on each rather than one chart for a cluttered multitude. Keeping a chart simple, however, does not mean that it has to be boring. Artistic creativity is welcome in visual aids, but the message must not be sacrificed.

Involve the Audience

Audience involvement implies establishing a rapport with your listeners so that each member becomes a willing recipient of your message. Audience involvement requires their familiarity with the subject and their interest in it. In Chapter 6 we urged you to take careful note of audience analysis. Knowing who your listeners are is the route to getting them involved.

Your audience must work with you as you go from visual or audio aid to interpretation and conclusion. Recall that in the learning process, you start with a given bank of knowledge and then you add the new. You go from the known to the unknown. A similar situation exists between you and your audience, and your visual aids must reflect that orientation. Objects and words used in illustrations must be familiar to the audience before the aid becomes effective. In a teaching situation, new concepts are introduced based on ideas with which you are already familiar. The learning process requires student involvement to move from the known to the new. This same approach is necessary for effective communication.

Walt Disney has been chosen to guide your judgment of a good visual and a poor one. The master of cartoon visuals used many excellent techniques that can be adapted to help you prepare effective supporting materials for your message.

TYPES OF AIDS FOR TECHNICAL PRESENTATIONS

Most technical presentations in the business world include communication aids. In this section we concentrate on visuals and discuss the most common ones: word and picture charts, objects and models, and photographs and films.

Presentational Charts

The most widely used visual aids for technical presentations are word charts with or without drawings. These charts are usually made into transparencies or 35mm slides that are projected on a screen. The masters can be hand-drawn, composed on office typewriters or word processors, created by office microcomputers, by minicomputer and mainframe computers, or by specialized professional services. Transparencies can be made from heat-sensitive film used with an office thermal copier or might be professionally made for a very formal presentation. The 35mm slides can also be made personally, through photographic interface equipment with computers running graphics software, or through professional suppliers.

Word Charts Word charts are an extremely important visual aid in industry. The simplest type used is the agenda chart, a straightforward, organized listing of the material to be presented. Such charts can be effective by themselves when you are speaking to your peers on a subject that is familiar to them. In this case, the chart serves as a public speaking outline, with all main transitional steps reflected on the screen. Figures 9-1 and 9-2 are two examples of word charts similar to most of the types used in presentations. Figure 9-1 is an agenda chart, detailing the organizational pattern the speakers will follow. Figure 9-2 is an example of a word chart that summarizes the main ideas of some part of the speech. Most industrial presentations would contain both kinds, along with complex charts integrating words and graphics. Together, these charts form the minimum that a technical business audience expects in visual aids.

The agenda chart was composed on a standard office typewriter using extra-large type. It is typical of the "create-your-own" chart used commonly for intracompany presentations. The idea chart was

AGENDA

June 11, 1985

INTRODUCTION/MEETING OBJECTIVE	PERSON A
TECHNOLOGY BACKGROUND	PERSON B
CURRENT PROJECTS	PERSON C
MARKET/PRODUCT SEGMENTS	PERSON D
COMPETITION	PERSON D
STRATEGY/APPROACH	ALL
CONTACT PLAN	PERSON A
WRAP-UP	

Figure 9-1 Simple agenda chart created on an office typewriter.

also prepared with an office typewriter. Very informal in-house presentations might use hand-drawn and hand-lettered charts for expediency. Any meeting of importance would make such crudely made materials unacceptable.

The common office typewriter or word processor is probably the most widely used means of producing charts. Many professionals prefer the mechanized neatness of a typed script even for informal sessions where the rough drafts are reviewed and a final copy must be

CONSIDERATIONS

- SYSTEM IS SOLID STATE, PRIMARILY PASSIVE
- STATE-OF-THE-ART TECHNOLOGY, NO INVENTIONS NEEDED
    - CONVERSION SYSTEM
    - HEAT SOURCE
    - COOLING SYSTEM
- IMPROVEMENTS POSSIBLE THROUGH
    - IMPROVED COOLING
        - BOILING
        - HEAT PIPES
    - HIGHER PERFORMANCE MATERIALS

Figure 9-2 Idea chart created on an office typewriter.

redone. Typewritten word charts are acceptable for communication outside of the corporation as well. They are used primarily in discussions with clients when work is in progress and revisions to the visuals are expected.

Standard typewritten charts are not generally used during a presentation to a diverse audience where the material reflects on the professionalism of the company. Such is the case when presentations are made to clients who will compare your company's work with that of a competitor. The use of professional-appearing visuals becomes a matter of persuasion and pride for your organization. Compare the three charts in Figure 9-3, one prepared by hand, one by typewriter, and one by professional equipment. Each chart has the same words but the message has a different effect on the audience.

What we have said about typewriters applies to word processors as well. By now, in fact, the word processor probably has replaced the typewriter in many offices. Word charts from one presentation are often appropriate for another and can be used repeatedly. Progress

Figure 9-3 A chart prepared three different ways—by hand, by typewriter, and by professional equipment.

— APPLICATION INVESTIGATION

PRODUCT MATCHED TO APPLICATION NEEDS

WORKING ENVIRONMENT

PRODUCT DESIGN AND FABRICATION REQUIREMENTS

— PRODUCT STANDARDIZATION

— TRANSITION FROM MEDIUM TO HIGH VOLUME PRODUCTION

DESIGNS THAT USE HIGH VOLUME TECHNIQUES

— CONTINUING MATERIAL IMPROVEMENTS

PROPERTIES

PROCESSES

MAJOR DEVELOPMENT ISSUES

-Application Investigation

Product Matched toApplication Needs

Working Environment

Product Design and Fabrication Requirements

- Product Standardization

- Transition from Medium to High Volume Production

Designs That Use High Volume Techniques

- Continuing Material Improvements

Properties

Processes

Figure 9-3 *(continued)*

and events may require small changes in the material on the charts. With a typewriter, the effort and cost of changes must be considered with respect to the audience before the changes are made; but the word processor has made it possible to update charts at very little cost so that improvements become practical.

Word processors are sometimes tied together through network links. Charts created in one location can be electronically transmitted to another for duplication using the word processor at the receiving location. This process can significantly improve the distribution of presentation materials to field sites. These mechanisms help to relieve the effort to obtain good and relevant visual aids. They offer

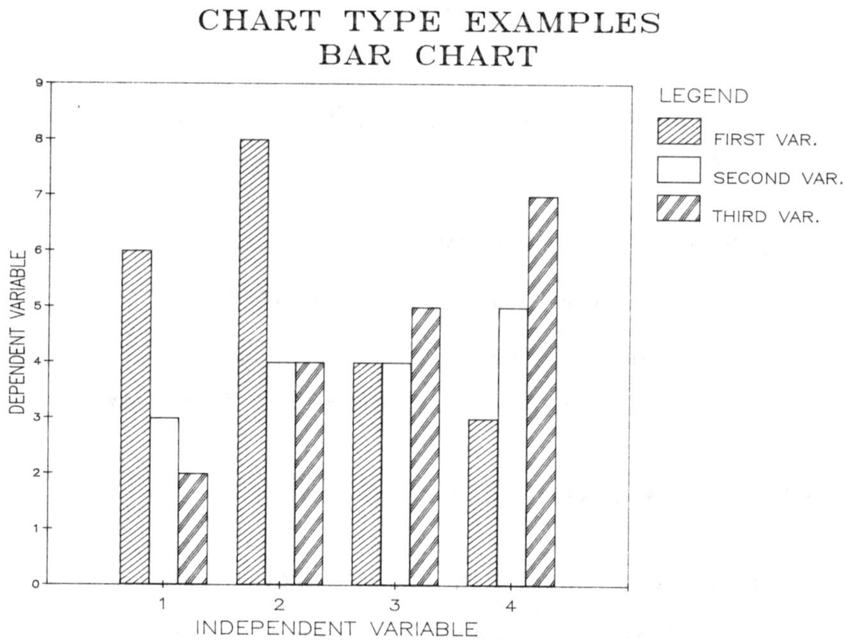

Figure 9-4 Chart type examples: bar chart, pie chart, line chart.

the opportunity for fresh and current materials in a short period of time.

Picture and Graph Charts A speaker should use pictures when they are available and appropriate. They add interest and are a contrast to the usual word charts that are part of most presentations. Pictures used in place of words reduce the chance for misinterpretation, and minimizing such entropy in oral communication is essential to the business world. Figure 9-4 shows three types of charts.

Picture charts take the form of drawings and sketches. Drawings are often used because they can be readily and simply obtained. Such pictorials are used in design reviews to describe the physical features and design details of a new product; they generally are created by designers as part of the normal flow of work on a project. As a result, they are usually readily available and can be made into transparencies. Less common are sketches that must be generated

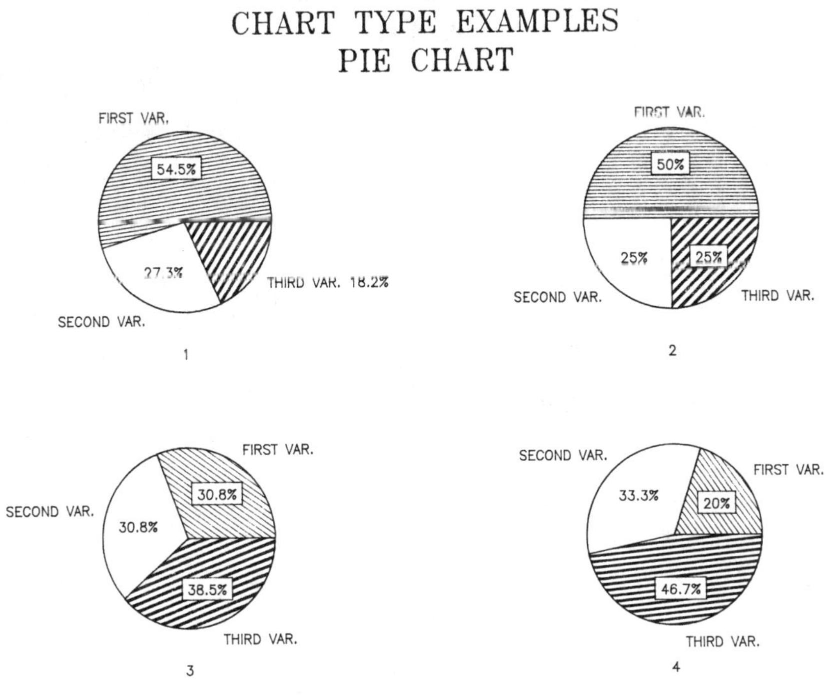

Figure 9-4 *(continued)*

just for the presentation. These are justified for customer or upper-management presentations.

Graph charts or complex charts containing graphs are usually used to show trends. Graphs take many forms, from pie charts and bar graphs to curves showing dependent-to-independent variable relationships. These charts help the audience understand the material and so they maintain interest in the presentation. Figure 9-4 shows a bar chart, pie chart, and line chart.

The new technology in office equipment and professional equipment simplifies production of picture and graph charts. Office copiers that reduce and expand are used to create standard-sized transparencies from oversized drawings. Drawings stored on microfilm can be blown up to the desired size. Personal office or home computers can

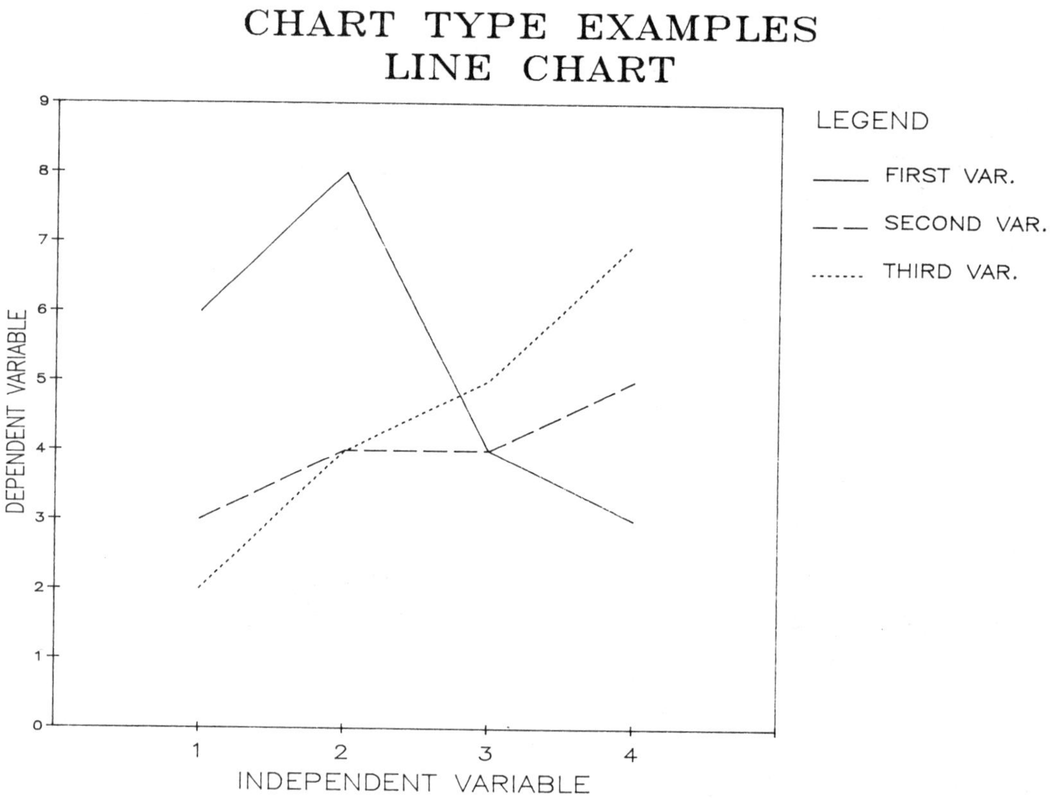

Figure 9-4 *(continued)*

be programmed to create graphics. Mainframe and minicomputers have long been used for graphics generation.

Computer graphics have improved rapidly in type and quality. An entire software industry was created with the advent of the office personal computer and desktop plotters. Shown in Figure 9-5 are two charts created with an early and then with an improved version of a software package written for a personal computer. In the improved version, letter-formation quality adds a professional appearance.

Professional-quality graphics using a computer are created from elaborate software systems written for micro- and minicomputers. These packages offer flexibility in format, color choices, and patterns, as well as a library of figures that can be used to create impressive graphics. All of these can be saved for later reuse or modification. The advantage of these systems is the high quality available and the short time needed to produce them. The chart is

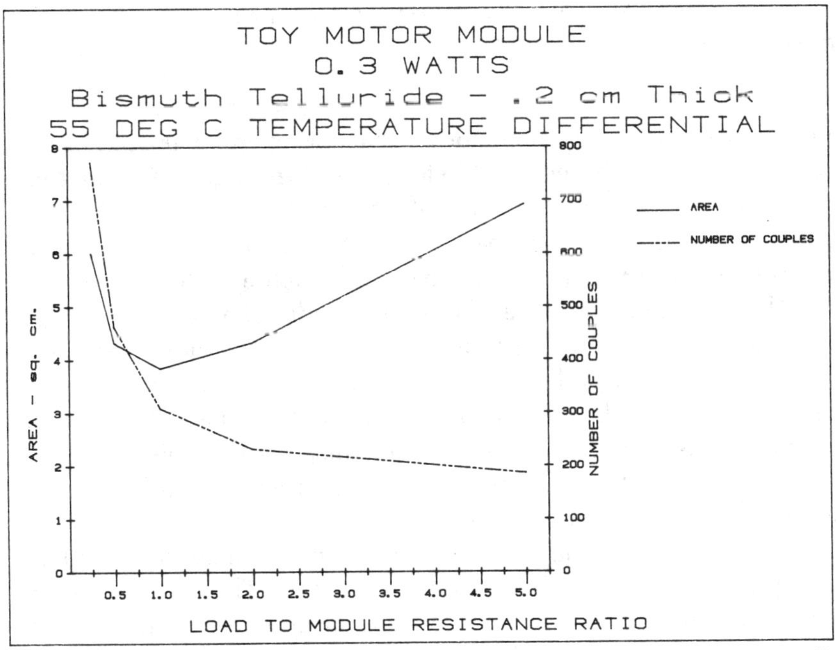

a

Figure 9-5 Charts created on a personal computer. Chart **a** was produced with an early version of the software; chart **b** was created with an updated version of the software. (Courtesy of General Electric Co.)

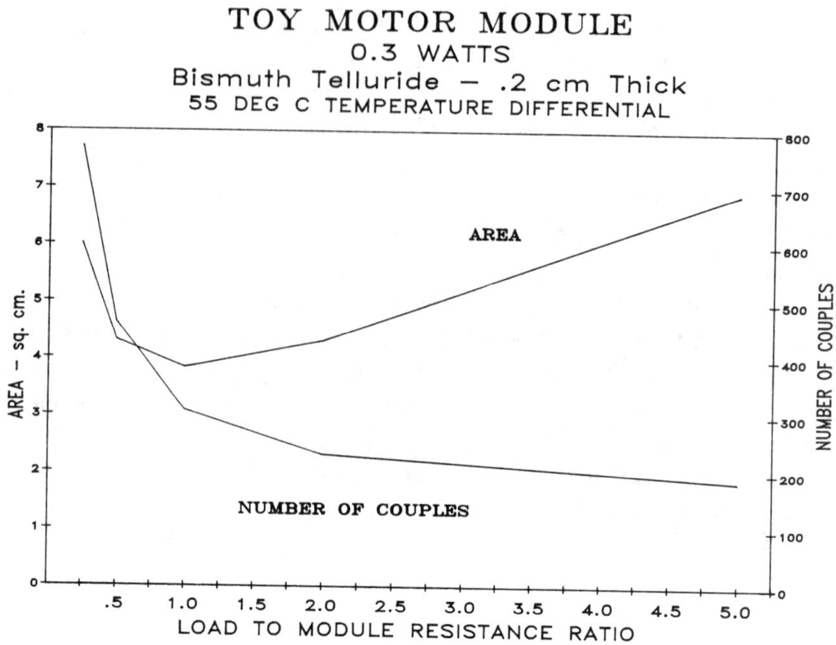

Figure 9-5 *(continued)*

prepared using a color monitor so that the operator obtains a true representation of the material before it is saved in the system. Outputs are generally in the form of 35mm slides that can be used directly or from which transparencies can be made. Figure 9-6 shows an example of a visual aid prepared with such a system.

Mainframe computers are used to provide graphics from TV signals generated by weather and earth-resources satellites. These sources have provided very high resolution graphics that serve as photos of the earth and clouds in wavelength bands for scientific evaluation as well as for visual portrayals. An example of a high-resolution portrayal of Baltimore obtained from a mainframe computer is shown in Figure 9-7 on page 200.

A subset of computer-generated graphics comes from systems that produce Computer-Aided Designs (CAD). These are minicomputers or microcomputers with software or architecture that creates complex graphics. The software allows the computer to create pictures of objects based on word commands specified by the operator. In many cases, graphics used in technical presentations can come directly from this source. In fact, the CAD system can manipulate the picture in many ways. Speakers can select the view that best serves

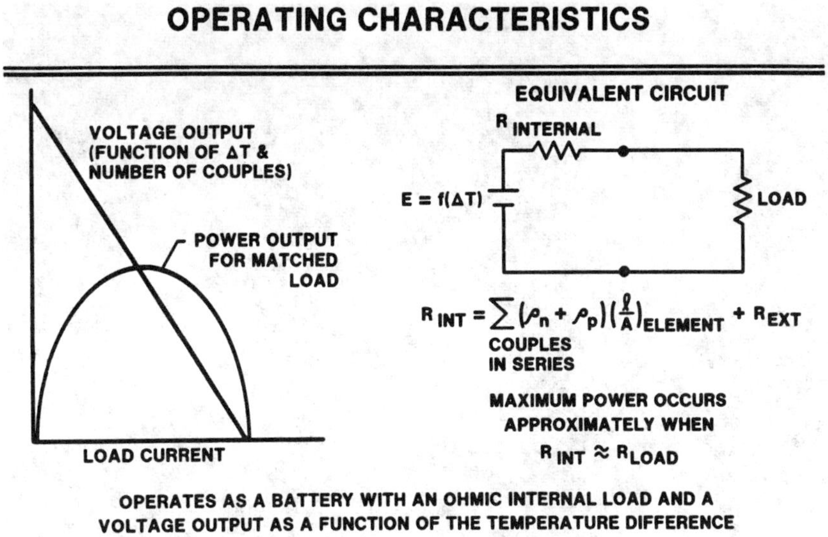

Figure 9-6 Visual aid prepared from dedicated professional system using a minicomputer. (Courtesy of General Electric Co.)

their needs. The picture can be scaled so that the whole or any portion of it is used.

Using Presentational Charts The effectiveness of any visual aid depends in part on how the speaker uses it. Slides and charts serve as separate channels of communication and have special requirements for their use. Slide presentations

- Require subdued lighting for best visibility.
- Can be manipulated by the speaker using remote control or by a helper advancing them by hand.
- Need the speaker's use of a pointer or light arrow to highlight important portions.
- Prescribe that the speaker stands free of the screen for maximum visibility.
- Make it easy for the speaker to fall into the trap of turning from the audience to see what is on the screen.
- Serve as great temptations for the audience to be distracted from the speaker.
- Must be arranged in careful order.

Figure 9-7 Landsat 4 TM image of Baltimore, Maryland, July 28, 1982. (Courtesy of General Electric Co., Earth Resources Applications, Lanham, Maryland.)

- Do not lend themselves to alterations before or during the presentation.

Transparencies, on the other hand, share some of the same constraints of slides but have different requirements for effectiveness. They

- Allow higher levels of room lighting, which maintains speaker presence to the audience.
- Still make it easy for the speaker to fall into the trap of turning from the audience to view the screen.
- Are usually much closer to the screen, allowing the speaker to stand by the projector. A pencil on the transparency may be substituted for a pointer or light arrow.

- Can be corrected with a grease pencil or transparency pens.
- Require the speaker to practice positioning them on the projector to avoid distractions from upside-down or backward placement.

Because of their many advantages, slides and transparencies are widely used during technical presentations. If used badly, they serve to increase the entropic losses; if used well, they enhance the message and its effectiveness.

Objects and Models

When their use is feasible, objects and models are extremely effective as presentation aids. The three-dimensional object is more accurate than any two-dimensional picture can be. Enhanced clarity is achieved when receivers can see and touch the actual article under discussion. All objects are not suitable for use in a presentation; and, in such cases, a model of greatly expanded or reduced scale is required. Following Disney's advice, you can design a model to distort reality by exaggerating size. For example, you could not use an actual microprocessor chip to show the circuit pattern contained within it; a large model or drawing would be necessary. On the other hand, the huge dimensions of a new aircraft under design would be scaled down into a model that fits on a desk top.

For both objects and models, cost is an important factor. Some models, for example, can be extremely expensive and must be used selectively. Whether object or model, poorly produced materials do more harm than good. Objects for use in presentations are usually created with extreme care. Models are generally one-of-a-kind and professionally fabricated to exacting specifications. Both objects and models are used for important presentations.

Using Objects and Models Unless the object or model is very large, their size constrains their use. Small objects simply cannot be seen much beyond the first audience row. Details in most models are lost unless the audience is free to examine them at close range. As a result, small group settings are usually best for such aids. The listeners are close to the materials and can examine the particulars.

You can invite members of a large audience to get a close look after the presentation has concluded. This procedure is obviously a compromise, for the force of the message will have diminished by then. The object or model, however, should not be left on full display during the speech because it will distract from the verbal message.

Many speakers delay the showing of these aids until late in their presentations, tying their conclusion together with the invitation to examine the model at close range.

Do not pass a model or object around the audience while you are speaking. The better your aid, the more seriously it will distract your audience from your words. Audiences sometimes begin to discuss the object among themselves while they handle it, and the speaker fades completely as a source of information.

Paper/Flip Charts

Transparencies and slides have replaced their paper predecessors. Paper charts are rarely used outside the most casual in-house meetings. They are difficult to prepare on the professional level and cumbersome to transport.

Paper charts can be used effectively in the small group settings. Because everyone is close to the speaker, the disadvantages are minimized. At an in-house casual meeting of a few persons, you can use them as "talking papers" to sketch ideas and generate lists. They are inexpensive and, in the right setting, help groups formulate wording for later refinement.

Using Paper/Flip Charts In case you do use these aids, you must be aware of certain problems. Take into account their size and visibility in the room where they will be used. Numbers and details fade from sight beyond the second or third row, so prepare them with care.

Make certain an easel is available for displaying the aid. Awkward propping or holding a large chart is extremely unprofessional. An aid that falls at the wrong time causes embarrassment.

The process of changing each chart requires that the speaker stand adjacent to the easel. This is the normal position if the chart is used to display action items that are checked off as completed during a progress-review meeting. The project leader would generally be the speaker and the aid would be a working document.

Photographs and Films

Photographs, like pictures and sketches, enhance the presentation and promote interest. In general, photographs are usually worked into a transparency or 35mm slide. Making a slide is simple, but having a photo put into transparency form usually requires the work of a media professional. For this reason, transparency photographs are used sparingly, being reserved for the special situation when the presentation justifies the cost of preparation. Figure 9-8 is an example of a photograph used by one of your authors.

Using Photographs and Film Films are seldom used as visual aids in the technical or business environment. One of the reasons is that a film ultimately takes the place of the speaker, who, in effect, becomes part of the audience in a darkened room and loses the vital contact that an in-person presentation requires.

However, films can be used in special situations. We recall a conference where a film of an aircraft, shown as a brief part of a presentation, was spectacular and supported the speaker's point. The overall message was enhanced. Film and its modern partner videotape have their place in presentations. They can support a message with unique impact but they are expensive and subject the speaker to an anonymous status during the showing.

Like objects and models, photographs should never be passed among the audience during a presentation. They are powerful visuals and will tempt a listener away from the speaker. A mounted photograph on display during a speech presents several problems. Unless

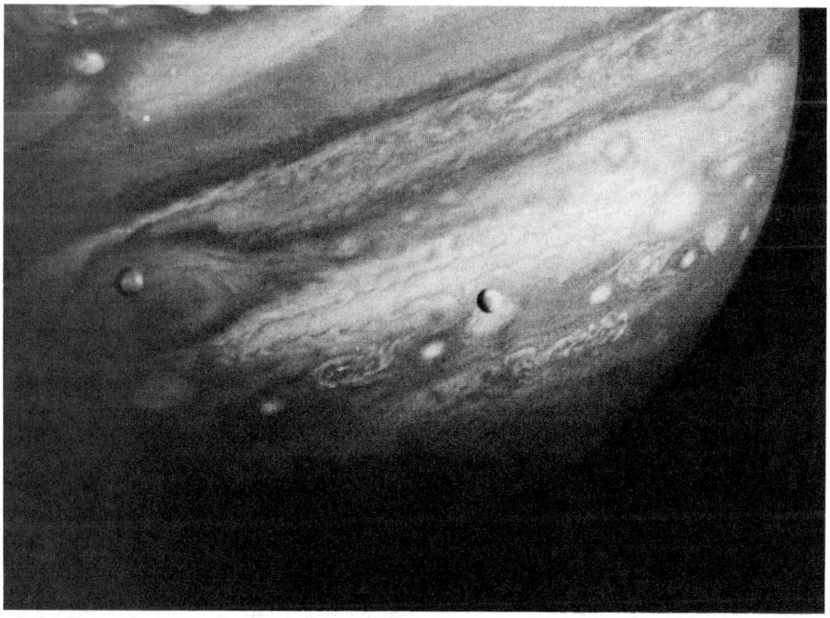

Figure 9-8 Photo of Jupiter taken by Voyager I. (NASA photograph Voyager 1-19, P-21082, February 21, 1979.)

its size is very large, it will not be seen. If it is large, portability becomes difficult.

Film needs a dark room for visibility. Under these conditions, a speaker becomes an observer or a disembodied voice. If film is to be used, it should be introduced and the main message verbalized while the room lights are up. The film clip is shown with the lights down. When the lights come up again, the speaker should reiterate the points made and how they relate to the message.

Both photographs and film require a professional's hand. They are difficult media. Poorly made home movies or amateur photographs detract from rather than aid the message. Fortunately, corporations often do have access to high-quality work, some of it created in-house. View anything you plan to use several times. If the aid can be integrated as a partial support of your message, use it. If it detracts from the message, choose something else.

COMMUNICATION EQUIPMENT IN MODERN CORPORATIONS

Modern offices are filled with communication aids. Familiar ones such as typewriters, word processors, standard telephones, and copiers need no explanation. A few are not so familiar or are used in novel ways. We have chosen to discuss several of these aids because they will probably occupy a large portion of your business day.

Telecommunications

Telecommunications consists of telephone system networks through which you can send oral messages, hard-copy messages, graphics, computer files, and telegrams. Telephones, telecopiers, remote computer terminals, as well as variously equipped microcomputers, form the basis of telecommunications systems.

Telephones The extremely familiar telephone forms the basis of the telecommunication system. Technical advancements have given this machine broad flexibility. Beside the usual single conversation between two parties, multiple hookups allow several parties in different locations to hold teleconferences. Such conference calls are extremely attractive financially because they save the cost of travel to a central location for face-to-face meetings. They also enhance information exchange by exposing all participants to the same voices, questions, rejoinders, and input.

Telecopiers A telecopier is an image-producing piece of equipment that uses telephone lines to transmit the image from one machine to another. A page of words, graphics, or both is fed into the sending machine. The image is converted into electronic signals and transmitted over telephone lines to a receiving machine. The receiving machine reconstructs the image from the signals received onto paper for use by the receiving party. This equipment permits near real-time transmission and minimizes the entropy involved in describing a graphic with words alone. While highly legible, the visual quality of the received material does not permit its use directly for presentational purposes.

Remote Computer Terminals

Through computer time-sharing, many users can access a central computer using a terminal. The terminal converts the keyboard strokes into a form that can be transmitted over telephone lines and received by the main computer. Responses from the main computer are transmitted in a coded signal back to the terminal, which usually translates it into hard copy. In this way, a single person at a remote terminal can communicate with many other people using the information stored in the mainframe system.

Microcomputers

This decade is witnessing the acceptance of the independent microcomputer into the business world. This versatile machine can be used directly in many office procedures. Equipped with a communication modem, it becomes a terminal capable of interfacing with a mainframe. With the modem, it can use the telephone network to transfer files from one machine to another far-distant machine.

Electronic Mail Terminals or microcomputers can also transmit material created on a computer to a central storage system. Receivers then have access to that material from their own computers. Sometimes equipment can be tied together to form networks where one computer can access the files of another directly without a separate intermediary storage system. In this way, a person can send a letter for storage in a central machine, and the letter can be read by a single receiver or anyone else who has the access code. This letter can be retrieved and stored as hard copy. Response to the letter can be sent, either for immediate or later retrieval. Such equipment makes the usual turnover time for sending and receiving letters through standard postal service seem antiquated.

This discussion of equipment is meant to introduce you to the diversity of approaches to communication aids that are already in place in the industrial setting. With the growth of technology, new mechanisms will replace the old with the objective of communicating better in less time and with less effort. You will learn to deal with this technology whether you seek to implement it first or whether you become the recipient of the communication.

WRAP-UP

Your career will require you to use communication equipment daily in the office. Giving presentations will be an integral part of your career. Visual and audio aids are central to technical communication. Your superiors and listeners will expect competence in handling these support materials.

Borrowing from Walt Disney, the master of visuals, we have presented some guidelines for creating good aids. We explored the machines of the modern office and a variety of presentational aids, some of which can be produced on these office machines. We focused on transparencies, primary among aids for the technical professional.

Finally, we suggested ways of using aids effectively. Regardless of your specific discipline, communication aids are necessary if you are to make your ideas clear to others.

ACTIVITIES

1. Present to your class a four-to-five minute explanation of some technical concept. Use at least three transparencies to help your audience understand and accept the concept.
2. Divide the class into groups of five members. Each group selects some technical concept it will explain to the class. Each member of the group will speak for two to three minutes using *one* of the following aids:

   - Word chart transparency
   - Picture and graph chart transparency
   - Objects and models

- Paper/flip chart
- Photograph/film

When the group has presented its explanation of the concept, analyze the different contributions each aid made to the class's understanding. Which aid would you conclude is the most effective?

3. Observe a classroom lecture on your campus where the teacher uses visual aids. Analyze:
   a. The quality of the aids used.
   b. The effectiveness of the teacher's use of those aids.
   c. How additional or different aids could have improved the class's comprehension of the subject.
4. Investigate your campus library as a repository of various ready-made aids. Prepare a handout for your class that lists:
   - Types of materials available.
   - Location of those materials within the library.
   - Procedures for borrowing the materials.

NOTES

1. Elena P. Zayas-Baya, "Instructional Media in the Total Language Picture," *International Journal of Instructional Media* 5 (1977–78): 145–50.
2. Paul Heckel, "Walt Disney and User-Oriented Software," *BYTE* 8, no. 12 (1983), 143–50.
3. Heckel, 144.
4. Heckel, 144.
5. Frank Thomas and Ollie Johnson, *Disney Animation, the Illusion of Life* (Abbeville Press, 1981), cited in Heckel, 144.

Summary: Part II
Preparation for Encounter

The central section of this book addresses preparation. We argue that neither simple answers nor formula shortcuts exist when a communicator sets out to get ready for a meeting with the self or another person. How a speaker prepares involves at least five separate considerations, treated here in the five sections of Part II.

Intrapersonal communication consists of the ongoing message exchanges a person has with the self. The Versions of Me model (Figure 5-1) depicts the ways people see themselves in these internal conversations and the ways they are, in turn, seen by others. These versions sometimes differ greatly, and self-concept is distorted.

Some of the messages people struggle with throughout their intrapersonal lives provide growth potential, and some are destructive. Each of us works through a set of values, decides on an appropriate ethic, and deals with the needs we feel. The conflicts between messages are argued and resolved in the privacy of intrapersonal communication.

Our fears and the ways we cope with them sometimes lead to miscommunications with ourselves. Speech communication fright, a widespread phenomenon, is based on one such erroneous message. Some people carry the fright beyond the context of public speaking to a fear of *all* communication. They cope by becoming quiet. This quietness, which might be akin to shyness, has serious effects on self-concept and successful relationships with another.

Even the most quiet of us reaches out to another person sometimes. We call this other *audience* and discuss the process of audience analysis. Although it is most obvious when the communication is in the formal setting, all speakers must consider their listeners from the aspect of who they are, what the speaker wants from them, and how they can be led to act. We take the position that communication, at heart, seeks some action: to satisfy the self, to motivate others, to incite others to act, or any combination of the three.

The corporate world, decidedly goal-centered, directs its professionals to four important goals:

- Acceptance of image
- Acceptance of ideas
- Acceptance of methodologies
- Acceptance of contracts

Underlying each corporate goal is the movement to ego-satisfaction, motivation, and action. Once the audience is known and the goals understood, the communicator is ready to select a topic and begin the research. Our suggestions for both indicate that sources are diverse and rich in potential.

We considered six possible patterns of speech organization. Which one or which combination is best for a given presentation depends on the audience and the speaker's purpose.

The content of presentations reflects the personal and corporate demand for proof. We conceive of proof as consisting of evidence and reasoning; the acceptance of proof is the listener's prerogative. Both evidence and reasoning used well bring acceptance of proof; misused, they contribute to speech entropy.

Finally, we discussed the way mechanized aids are used in technical communication. Such aids are not only important to the clarification of messages, but they have also come to be expected in the corporate world. Ranging from commonplace tools such as office typewriters to the complex computer-generated visuals, they are devices that technical professionals must know how to use effectively. Speakers must master the guidelines for using aids, whether their presentations will be in-house or to people outside the parent company.

With preparation now completed, communicators move toward the time when they actually meet their audience. All of the tools are in place. Now they must be assembled and brought to life in the four separate settings that are the subject of Part III.

PART III

SETTINGS FOR TECHNICAL COMMUNICATION

CHAPTER TEN

ONE-TO-ONE: DYADS

Underlying the various concepts we have discussed throughout this book has been our assumption that *all* communication is dyadic. The basic model, presented in Chapter 2, visualizes communication as happening between *a* sender and *a* receiver. When that sender and receiver are the same person, we called the interaction intrapersonal. As social creatures, however, humans use their symbol systems to contact others, and the communication is always completed on a onc-to-onc basis.

Many communications certainly take place when there are more than two persons present. From a small social group to an audience of millions, a sender often has multiple receivers. Yet, the message is created by *each* of these receivers working uniquely on the same stimuli that the sender transmits. In this sense, communication is always either intrapersonal (the sender and receiver are the same person) or dyadic (the sender and receiver are two separate persons).

In this chapter we address the special factors involved when only two persons are interacting. Our focus is the way dyads function in the business world. Specifically, we discuss the formal interviewing process, the interactions of coworkers who are superiors and/or subordinates, the problems of conflict management, and the realities of giving or taking orders. In each case, we consider these situations only when they involve no more than two different persons.

INTERVIEWING: THE INITIAL DYAD

Even a brief scan of magazine articles, bestsellers, and philosophy books would convince you that each human desires contact with another person. Psychotherapist Rollo May says, "For human beings, the more powerful need is not for sex per se but for relationship, intimacy, acceptance, and affirmation."[1] In *Bridges Not Walls*, John Stewart makes the same point when he says that "the quality of each person's life is directly linked to the quality of the communication he or she experiences."[2] If May and Stewart are correct, then a significant portion of life's quality stems from dyads developed in the professional setting. We do not downplay the importance of either intimate or casual meetings, but professional life demands mobility. With family and childhood friends left behind, a professional's business associates become extremely important. We need to look at how these associates interact. Specifically, we will discuss how interviews begin careers, how one-on-one conversations further them, and how the giving and taking of orders constitute them.

Most students have had some interviewing experience before they reach the end of their college days. They might have been interviewed by an academic department before being admitted to a given major, by the student paper seeking answers from "the student in the street," or by an employer before being hired for a part-time or summer job. These experiences are very helpful, for most of us are extremely anxious as the time for that first postdegree professional interview approaches. In this section, we look at the job-interviewing process and provide some information to help you deal with the anxiety involved.

An *interview* is defined as an in-person communication entered into to accomplish some specific task goal and revolving around specific information. This type of communication usually involves more questions and answers than others do. This definition includes many kinds of interviews: counseling, disciplining, evaluating, medical/legal/data information gathering, survey taking, opinion gathering by mass media, or job screening. Some are conducted long distance, via written questionnaires or phone surveys. Some are also done by panels, such as a candidate selection committee. We restrict our consideration here to those between two people, in-person, and for the purpose of securing employment and/or promotion. Further, our focus is on the person *being* interviewed.

Interview Preparation

Preparation is the first step in this communication encounter, just as it is the first step for all communication. Everything we have said in Parts I and II of this book now comes into play. Your goal is persuasion to action: you want the interviewer, motivated by your messages, to offer the position to you rather than to other candidates.

Before you step into the interviewer's office, considerable communication will already have taken place, and both of you will have done much preparatory work as well. This communication involves such things as:

- The job announcement that you saw or heard in some form.
- Your résumé, carefully prepared and sent with a cover letter.
- A follow-up phone call you make a few days after the résumé should have been received.
- The phone call or letter from the interviewer requesting additional information or an appointment for an in-person meeting.
- The phone call or letter from you requesting information regarding the company and confirming the time of the interview.

Once the interview has been arranged, you must prepare carefully. Decide how you will present yourself physically: neat appearance, carefully chosen clothing, shined shoes, and careful grooming. As we said in Chapter 3, these nonverbals send important messages, and you must decide what you want to say about yourself when you use them.

Collect any documents you think the interviewer might wish to see. Such things as social security and driver's license numbers and other pertinent licenses and certifications must be readily at hand. If your résumé did not include your letters of recommendation, make sure you have copies to leave with the interviewer. Such letters should have been secured long before you began job searching. Addresses and phone numbers of previous employers should be available for the interviewer.

Research the company that you are visiting. Few things strike an interviewer as more incompetent than does a job candidate who knows nothing about the history and organization of the company. Stock reports and annual reports for large corporations are matters of public record. Read them. For small firms, get information directly from the firm itself. Arriving a few minutes early allows you to read any information usually scattered around waiting rooms.

You do not always know in advance who will be doing the interviewing. This is that one-size-fits-all risk category of audience analysis. Prepare for a generic interviewer who is searching for someone who fits the job description and objectives of the particular company. If you know who the interviewer will be, your task is simplified. Your audience analysis can be specific and your message tailored with some precision. Do some hard thinking, in either case. If your interviewer is representative of the company, and you find that this person disturbs you, perhaps the job would disturb you also. If you think you are not the kind of person the company is looking for, discover this as soon as possible. Serious preparation and audience analysis is necessary before you can consider these questions.

Preparation for an interview must consist of planning your answers to those questions that surely will be asked. Consider what questions you have for the interviewer as well. Remember that, under the stress of the actual event, impressive answers and perceptive questions may not occur to you.

A good preparation step is to reread all the materials you have submitted. What else do you think the interviewer will want to know? What points need elaboration? What negative factors require explanation? What added materials do you plan to discuss? What things about the company, the position, the salary, or the department do you wish to know? These questions are probably the commonplaces of interviews, and serious thinking is required to handle them.

Questions and Answers

Questions and answers form the heart of this dyadic encounter. The interviewer's purpose is to ask for information that will serve as a basis for forming an impression of you. Your purpose is to provide the information in such a way that the impression leads to a job offer.

The give-and-take of questions and answers should not mislead you. *Many* other kinds of messages are being exchanged at the same time. Seemingly idle conversation is almost always an icebreaker. Chatting about your campus, the latest sporting tournament, or your most recent position are ways of putting you at ease. The nonverbals of *how* you present yourself are apparent too. The candidate whose words say "poised" might be giving away inner fear by the kinesics of twisting and shifting postures. Downs, Smeyak, and Martin remind us that "Research has emphasized the relative importance of the visual cues and suggests that when people receive contradictory messages, they are more likely to believe the visual ones."[3] Make certain that your nonverbals are sending a positive message.

With these multiple symbol systems in mind, focus on the questioning process. The questions you are asked will be either closed or open-ended. Neither necessarily gives better information about a candidate's qualifications, so both or variations of both are used frequently.[4]

A closed question requires a specific piece of information and restricts the response. Examples of closed questions that you or the interviewer will probably ask include:

- Who was your supervisor on the Zeno Project?
- How long did you remain in the accounting major before declaring the major in earth science?
- When did you travel to Japan?
- What is the salary range of a level-two engineer?

Open-ended questions, on the other hand, permit a range of unrestricted responses. They might include:

- What do you hope to get out of this job with us?
- Why did you leave your last position?
- What made you decide to switch from accounting to earth science?
- Why is there a two-year gap between college and your first job?

Alan Garner suggests some key words that give you a clue as to which kind of question is being asked:[5]

Closed	*Both*	*Open-Ended*
Are?	What?	How?
Do?		Why?
Who?		In what way?
When?		
Where?		
Which?		

Each kind of question has advantages and disadvantages. Closed questions are efficient, but usually do not allow much depth of response; open-ended questions allow a person to display personal qualities quite revealingly, but they are extremely time-consuming. The more demanding the position sought, the more likely that open-ended questions will figure prominently.

Keep the following list in mind regardless of the type of question being asked:

- Plan ahead. Prepare. Prepare.
- Present yourself in the best possible light, but never sacrifice honesty. Lying to an interviewer is a question of ethics, but it also invites irreparable damage to your credibility.
- Make certain you understand the question. Use interviewer feedback to double-check the accuracy of answers.
- Elaborate on answers, even to closed questions, if a simple answer will not give a fair picture.
- Do not ramble. Interviewers are busy people who appreciate direct and well-considered answers.
- Be diplomatic in your descriptions of other persons or employers. Story telling and name-calling tend to cause backlash.
- Form your own questions clearly enough to secure the information you seek.
- Do not fear silences. Interviewers sometimes use the pause to encourage a person to keep talking, thereby saying things better left unsaid.[6]
- Remember that this is a communicative act, and your communication skills are being measured too.

Interview Courtesies

Manners and interview behavior are more than just "being polite." They communicate social competence and, as with all other symbol systems, contribute to the message an interviewer decodes. Many of these courtesies are nonverbals, and none should be underestimated.

You are expected to be on time, having secured adequate directions to make sure of your arrival. Hearing "I got lost" might elicit sympathy from the interviewer, but not much respect.

If you are invited to lunch (often the case during a job hunt) follow your interviewer's lead regarding how elaborate this lunch is to be. Your host or hostess might envision a light courtesy meal, while your expansive order makes it something quite different. Experienced interviewers are usually sensitive to the candidate's uncertainties and try to clarify the expectations. A statement such as "They have the best veal in town here" might be saying, "Go ahead; order the veal. It's the most expensive and elaborate item on the menu, but we have the resources for it."

Attempt to show your interviewer that you are able to deal with your own anxieties. These people try to make applicants feel at ease. Communicate their success nonverbally as you begin to relax. You may even enjoy the experience.

Never insult your interviewer by appearing ungroomed and unkempt. If you are unaware of the expectations concerning dress and neatness, consult your university placement service for advice.

Follow your interviewer's lead regarding the amount of time allotted to the entire appointment. Your long stories or detailed explanations may be encroaching on time set aside for something else. Allow the interviewer's nonverbals to tell you when the appointment is nearing its end.

Always follow up the interview with a thank-you letter. This simple act of courtesy is extremely valuable. It not only tells the interviewer that the time spent has been appreciated, but it also tends to remind the interviewer of who you are. Remember, several candidates will be interviewed, and memories of the impression you made might grow dim. Help the interviewer remember you with a thoughtful note that cements a positive memory.

From its planning through to the actual encounter and its conclusion, an interview is the first career trial of a young professional. Except in the most unusual circumstances, every career begins with one. If you keep in mind that it is a communicative act, persuasive in nature and bound by the special expectations of such dyadic encounters, it will start you on your way.

CAREER DEVELOPMENT: MAINTENANCE DYADS

From your first day on the job, you form many important relationships. The migratory nature of many professional jobs makes the work environment a source of friendship, as well as a place to develop a business reputation.

Interpersonal Relations

Interpersonal relations are the associations built among dyads. When they are a part of your professional life, they take on added layers of importance. Some people with whom you interact daily remain only casual associates. Neither of you desires to expand the encounter beyond the working day, but you both can count on seeing each other sometime during business hours.

Some people move beyond the casual encounter to a cooperative task. Co-working groups form for the duration of a project, only to disband at its completion. From these groups grow reputations: this geologist is a good person to have in a tough spot, or that chemist knows how to ease tensions when the pressure becomes overwhelming. Such impressions are formed from the communications of individual dyads over time.

A few dyads grow beyond the work setting into friendships. Many different factors cause people to find each other mutually satisfying and begin to speak of another person as "friend." Such relationships often extend to activities having nothing to do with the business setting. So it is that a casual discussion about a mutual love of tennis results in a Saturday morning tennis game. The game may become a steady activity and eventually leads to other joint activities.

Sometimes, even true intimacy might develop with one who is first called "friend." As with all intimate relationships, a point is reached where the other person takes on a special status. The exchange of private information leads to trust between the two people.

One of the difficulties women face in the professional world is their uncertainty about how others will see relationships. Friendships across gender are hard to establish because suspicions seem to rise at once that sexual intimacy is taking place. Many women, especially those in the traditionally male-dominated disciplines, solve the problem by avoiding friendships with coworkers outside of business hours. Such a choice is difficult, for these employees rob themselves of relationships that are taken for granted in single-gender dyads.

Networks

A *network* is usually an informal communication system that uses personal dyads in series to transmit information. If some message such as "Bob is looking for a job" needs to be sent, one member of the network calls another, who calls another, and so forth. The message is "broadcast," sometimes very widely, and Bob's chances of finding a new position thereby increase.

Networks are an effective means of sending and securing information. In Chapter 7, you will recall, we listed coworkers as an excellent source of research information. They also might be political sources. In some organizations, only accepted network members are considered part of the inner circle. You can imagine the potential conflict when several such networks exist in the same corporation, each in competition with the others.

Entrance into networks is through dyads built step by step. Two persons impress each other, so the one introduces the other to a few acquaintances. These, in turn, are impressed and a reputation for being "okay" is built. If Bob is one of these okay persons, he becomes part of a network whose members are willing to call other people who do not know him in order to ease his way. Bob, of course, is also expected to provide the same information or introduction for others in the network.

Women, especially those taking the safe path of not establishing friendships with coworkers, often find themselves left out of networks. Thus, Jane finds out about the vacancy the day *after* Bob fills it. In disciplines where the gender distribution is fairly normal, women often form their own networks. In corporations or disciplines where few women are represented, women face serious obstacles.

Daily interactions structure how any person, man or woman, is going to be judged. Arrogance, inability to take part in required activities, refusal to interact, snide comments, or other negative dyadic interactions will isolate you. When you need to rely on a network of persons for help, you might find that no one is willing to activate the process.

Conflict Management

Managing conflict is made necessary by the entropic losses in communication. The essential differences between people—who they are, how they think, what they value, what they desire—all result in messages being decoded somewhat differently by different receivers. Human beings are unique, never-to-be-duplicated individuals for whom, quite literally, nature throws away the mold each time a child is born. In this uniqueness is the source of communication loss as well as the potential for conflict.

We define *conflict* as a *self-recognized communicated difference* between people. That is, one or both of the dyadic members is conscious of opposition. Entropic losses occur continually, but until they are *recognized* by at least one participant, conflict has not yet occurred.

Since entropic losses are inherent in communication, potential for conflict is always present. Nevertheless, not all dyadic relationships are doomed to deterioration. Conflict can be a positive thing. Ideas can be expressed, values examined, and hard choices made. In the process of examination, both people are able to understand each other and themselves better. The dyad can actually strengthen.

Neither is it necessary for conflict to become unpleasant. People finding themselves in opposition on some point can agree to disagree. For some people amiable disagreement is difficult. They find it hard to accept that a coworker, let alone a special other, has a different viewpoint. They want to "prove" to the receiver that *their version* is *exactly* the right version, or they want to make *all* the differences *absolutely* disappear. It cannot be done, any more than you can pour every bit of a liquid into another container. The losses are there in either case; the best you can do is minimize them.

Conflict management is about minimizing losses before they throw an entire relationship into hopeless disarray. A first step in managing conflict, however, is to realize its different sources.

Conflict due to physical impossibilities is a common problem. Two activities happen at once, and a single person cannot handle both. Your boss says you must be in Iowa on the fifteenth, and that is the day your good friend is getting married in Seattle. The ski club is having its Gala Weekend at the resort, and a major lab report is due on Monday. Most of us have had enough of such conflicts to make us wish for a clone. They cause considerable interpersonal as well as intrapersonal conflict.

Verderber and Verderber suggest that some "physically impossible" conflicts may be "pseudoconflict."[7] That is, with some compromise, you can meet both commitments. You need not stay the entire weekend with the ski club, for example. Doing some skiing and some report writing allows you to accomplish both goals. The conflict stems from a misunderstanding: the activities are not mutually exclusive.

Missing your friend's wedding involves quite another sort of physical impossibility. Here the activities do involve hard choices, because you cannot do both. A compromise of visiting the newlyweds next month might soothe any hard feelings, but their wedding date forces you to choose between your duty to your job and your affection for your friend.

Conflict due to differing interpretations means that one person in the dyad understands a message to mean something different from what the other person thinks it means. It may be the interpretation of data, the project goal, the correct way to write the proposal summary, the real intent of the supervisor's memo, or the interpretation of customer specifications. Whatever is at issue, its meaning is translated in two different ways.

When checking on "what that *really* means," people need to determine how far apart their versions actually are. One of us recalls a discussion between two technicians that was rapidly growing heated. They were at conflict over what some written instructions for the revision of several drawings meant. The one insisted that the entire set had to be redrawn, while the other claimed that the memo only criticized two drawings in the set. The solution was fairly obvious: call the supervising engineer and find out what he wanted. As it turned out, neither technician translated the engineer's message accurately. The portion needing revision involved views rendered in six of the drawings, roughly half the entire set. Material conflict is best handled through checking. If defensiveness about who was right and who was wrong can be kept to a minimum, the working relationship can continue unharmed.

Conflict due to methodology is a persistent problem in the technical disciplines. We have already spoken at length about the place of method in scientific procedures. If methods differ, so will opinions about which one is the right one for a particular problem. Often everyone agrees that something must be done, but *what* that something should be differs significantly.

The discipline itself usually offers some guidance concerning the methodology to choose. The program managers must determine what is acceptable practice, what precedents have been set in the past, how deviations from precedents have been proved. Again, if people become defensive about a certain method, they will be unable to assess accurately how receivers are likely to view it. Those who oppose their "right way" are also likely to be seen as opponents, and the interpersonal relationship is in danger.

Conflict due to ego injury is difficult to handle. People become identified with their ideas; and, when the idea is rejected, they feel as if they themselves have been injured. It is easy to advise, "Separate yourself from your idea," but most of us do not. The idea becomes us, and when the project manager dismisses our plan as ridiculous, we feel as if we have been labeled ridiculous as well.

Managers must be sensitive to the identification their subordinates develop with their projects. Decisions demand choices, so one plan will have to be chosen over some other; but the managers have to remember the people behind the plans. The person whose plan is not chosen must be reassured that the work done was excellent, that the person's competence was never in question, and that, as a token of trust, another valued assignment will be made at the first opportunity. Managers must also remember that their own egos are involved when pet projects come under attack. Whether with the self or another, ego is a persistent source of conflict.

Sometimes the ego conflict shows itself in struggles for power. We shall be speaking of power structures and how to handle them in the rest of this chapter. For now, remember that conflict often surfaces because the ego-picture one person has of his or her own competence or position, is threatened.

Conflict due to personal values surfaces often in business and personal life. In Chapter 5, we linked the formation of a person's values to self-concept. When somebody questions our values through communicated conflict, our actual "Versions of Me" seem threatened (Figure 5-1). A supervisor might have to fire a subordinate for repeated carelessness, for example. The general manager says, "Make sure that person never gets a job in this industry again." The supervisor's values, however, do not include a Version of Me as a person who totally destroys another's career. That supervisor's values are in conflict with those of the manager.

Because values are such a personal component of the Versions of Me, they are not easily changed. The supervisor wants to fire the worker while maintaining the worker's ability to get another job; the manager wants to fire the worker while destroying any future employment; the worker, by the way, has any number of explanations why other priorities were chosen over careful work. Each has personal Versions of Me at odds with every other version. Such a complex picture is not easily clarified.

Conflicts over values are probably best handled as opportunities to understand the other's version of the self and to explain your own. You are not going to change the other version, but you can agree to note what the other believes. The manager can learn that this particular supervisor will not operate in a vengeful manner. The supervisor can understand the manager's style also. They will not change each other, yet the worker must be dismissed. The supervisor might tell the worker that the work warrants complete destruction of career and, further, that no effort will be made to lie about the irresponsibility to future employers, but that no point will be made to publicize it either. If the manager is not satisfied with this action, then the manager should do the firing. Neither view is necessarily wrong, except to the person with other values. If both can see this difference and tolerate it by agreeing to disagree, the task of firing will get done and the relationship between the manager and the supervisor can continue.

Coercion to act contrary to personal values is extremely damaging. For example, the manager could force the supervisor to write letters damning the worker. The supervisor would have to act against self-image and, of course, would feel used and bitter. "It's not my fault. I would never do a thing like that, but that unprincipled manager forced me." An ongoing working relationship between these people becomes difficult.

Handling Conflict How a person handles conflicts depends on many variables: versions of the self, versions of the other, price extracted for the solution, environment where the difference occurs, and perception of coercion. Verderber and Verderber suggest that there are patterns of behavior people use to manage their conflicts.[8] They name five, and we have added a sixth:

- Withdrawal
- Surrender
- Aggression
- Discussion
- Persuasion
- Verification

We have added verification to their list to take into account the frequency with which technical professionals have to check accuracy. Interpretations of message content need to be made in all human dyads. Whether the train leaves at 5:00 P.M. or 5:10 P.M., whether she said to meet at the corner of Eighth and State Street or Elm and Tait Street are not unusual sources of conflict. Technical professionals, however, work in environments where exactness is part of the very method of science. Conflict over content must be anticipated; verification provides a way of dealing with it.

Each of us probably tends to favor one or another of the preceding patterns of behavior when faced with conflict. Whether we run from it in withdrawal, surrender to the other in false agreement, strike out physically (or verbally) in aggression, sort out viewpoints through discussion, persuade the other person to change, or verify assumptions, we are attempting to manage the differences we have with another. Some of these managerial actions are obviously more positive than others. Professionals who run from every controversy are no more effective managers than those who strike out harshly. Each of us, true to our own version of self, must develop sets of skills to help understand and resolve conflicts. The differences between people are always there; recognition of those differences makes them conflict. Conflict, in turn, can be an opportunity to understand the other, to understand the self, and to allow both to grow.

POWER STRUCTURES: GIVE-AND-TAKE DYADS

Just as we have difficulty recognizing conflict as a potentially positive force, so we become uneasy when we consider power. Pictures of potbellied bosses browbeating helpless workers blur the realization that all dyads involve power. It may be equal, unequal at different times, or vastly unequal at all times, but every relationship with another person is based on *one person's ability to get another to do his or her will and the other's granting the person's right to do it.* Cooperation is not at all denied in this concept. You and I might work well together in dyad on some common concern and do it as peers. Notice, however, the interaction implicit in this definition. One person communicates, exercising influence to get the other to do something, *and* the other allows that influence to operate. A few seconds later the roles are reversed with the former influencer now being the one influenced. We define *power* as the ability to influence an action or thought of another.

Reporting on a study of 754 adults in managerial positions, Andrew DuBrin summarizes the way each maintained influence and gained "mild degrees of power" in day-to-day encounters as:

- Assertiveness (being forthright with your demands)
- Ingratiation (getting somebody else to like you)
- Rationality (appealing to reason)
- Sanctions (using threats of punishments)
- Exchange (performing reciprocal favors)
- Upward appeal (asking for help from a higher authority)
- Blocking (engaging in work slowdowns or the threat thereof)
- Coalitions (getting the support of others)[9]

None of these is an unfamiliar tactic, although some are surely more positive than others. All are means people use to gain influence over others. DuBrin goes on to say that, when managers were asked to name the personal characteristics of the most effective influence-users, they listed "articulate" and "sensitive" as the top two.[10]

The power to alter another's thought or behavior might come from several sources. Such a person might

- be the one acting in some official capacity, such as manager or lead technician.
- be the one who has certain knowledge or experience that another wishes, such as a teacher.
- be the one held in esteem or affection by the person being influenced, such as a friend or spouse.
- be the one with a coercive retribution as a consequence of not complying with the action desired, such as an evaluator who has the right to fire.

Any such person can influence the thought or action of another. In our definition, the person doing the influencing has *power*. The give-and-take of influence is the basic constituent of all relationships. We intend to make you one of those articulate and sensitive influence-wielders, whatever your source of power might be. We must now look at the power structure of dyads, how to recognize their effective functioning, and how to improve communication within them.

Power and Superior-Subordinate Interaction

We will call the person *having* the influence to affect another's actions the superior, and the one *being* influenced to act the subordinate. These terms do not have qualitative connotations, do not imply better or worse. They are meant to describe the direction of action. Neither does it necessarily matter who is titled the superior. The one with the title Program Manager might get the other to do something as the result of a given communication, but then again, perhaps not. The environment of the testing lab probably makes the manager's words result in actions; but, out in the parking lot, when the same two persons are discussing what should be done about a car that will not start, the roles might shift. The environment of the home presupposes the superior status of a parent to get a child to obey; but, when the child is showing the parent how to work a personal computer, the one doing the influencing changes. Each communication has to be examined to learn who is doing what at which particular time. When a dyad is observed, the flow of communication shifts from one to another as one attempts to secure the other's compliance.

Viewed in this way, power structures in personal and business relationships become a matter of communication analysis. Further, they are not static. In situation A, the superior-subordinate roles might function in a certain power direction. In situation B, they could reverse. They might even alter within the *same* situation. The parent learning about personal computers from the child might be comfortable in the subordinate role as long as it involves orders such as "Type in the second number" or "Use the delete key." When the parent says, "Okay. That's enough for tonight. Time for you to go to bed," the usual superior position has been reestablished.

We will discuss a great deal of information on the subject of power by using the summary that follows. We contend that each person in a dyad, regardless of where and when the relationship occurs, needs to be aware of how the superior-subordinate roles function. To discover how the roles work, we have identified five concerns about power and the implications of each to both participants in any dyad.

1. Understanding the job description/role expectation

Superior	Subordinate
a. *What does it entail?* The total job/expectation of the role must be clear. Be willing to discuss and explain.	a. *Same* Clarity might require probing. "Is this my job/your expectation of me?"
b. *What are its responsibilities?* The duties and accountability should be understood.	b. *Same,* but understand which one of you has final accountability for any project. Question differences in perception.
c. *What are its limits?* At what point the power stops needs clarification.	c. *Same* Your rights to protection must be established.

d. *What are the interpersonal concerns of the position?* The roles of superior and subordinate communication patterns are set down by the culture. Use them with tact.	d. *Same* Patterns that the culture establishes for communication with superiors come into play.

2. Understanding the corporate/personal expectation for output

Superior	Subordinate
a. *What are these expectations?* Output refers to a particular project. Expectations are sometimes written and sometimes understood informally.	a. *Same* Make certain you do understand. Ask to check your assumptions.
b. *How does output responsibility overlap with that of others?* Conflict potentially exists in areas of ambiguous overlap.	b. *Same* Subordinate overlap is quite possible. Know who has the power to resolve problems.
c. *What are the consequences of falling short of output?* Understand the rules, sets of procedures, conditions for, and limits of punishment.	c. *Same* Protections against intimidations must be understood and used.

3. Understanding the chain of command

Superior	Subordinate
a. *How is the power chain structured?* The formal organizational chart of power does not have to be the same as the informal one in operation. Know which is which.	a. *Same* Know where your immediate supervisor fits in the chain.

b. *Is the chain of command limited by any special constraints?*	b. *Same*
Total control does not exist. Know where the limits are. Ask, "Is this my decision to make"?	Nobody has the right to control those parts of your life that lie beyond the limit of granted power.
c. *Where do I want to be in the power chain?*	c. *Same*
Examine carefully what each level of command will entail and decide if that is what seems comfortable for your talent and personality.	Examine realistically where you wish to be in the command chain. Freedom from control is not possible, but which level would be right for you?

4. Understanding the way communication flows

Superior	Subordinate
a. *What kind of channels are most prevalent?*	a. *Same*
Knowing what kind of channels are used most frequently in such dyads allows for monitoring of special problems and developing of special skills needed for those channels.	Evaluation will take place to a large degree based on the channels used most frequently.
b. *Is feedback allowed to function freely?*	b. *Same*
Defensiveness is a block to honest decoding of what is being said.	Determine if the other truly wants honest feedback.
c. *Is listening recognized as different from hearing?*	c. *Same*
Seek out the errors due to poor listening. They are costly mistakes. Train self and others to decode well.	Ask for training in decoding skills. Errors are the fault of *all* participants in any communication.

Superior	Subordinate
d. *Is open communication, free of threat, the norm?* Threat engenders fear, and fear closes off accurate communication. Erroneous communication, in turn, results in mistakes and low morale.	d. *Same* Closed communication will be fearful and frustrating. Seek out why your partner will not open the lines to an honest exchange.

5. Understanding personal ego

Superior	Subordinate
a. *Where are the personal danger spots?* Identify the issues, nonverbals, challenges, and situations that set off your personal intense feelings.	a. *Same* Know how you feel when another has the right to direct your actions.
b. *What options are usually taken to manage conflict?* At least six ways of handling conflict exist. Recognizing which are selected most often reveals much about the self.	b. *Same* Verbal and physical violence cannot be allowed to develop into a persistent pattern. Protections must be invoked.
c. *What kinds of protections to the ego are most often used?* Injured egoes need healing. How one goes about the process is important to the self and the other.	c. *Same* Allow feelings to be soothed. Blame and bitterness do more harm than good.

These processes of analysis are certainly not easy. They dovetail and enforce each other. None is static. We all function as subordinates in some relationships, and each of us serves as superior in others. In turn, the relationships alter constantly. The person doing the influencing one minute is being influenced the next. We see it as an error to assume that all communications with a titled "boss" necessarily mean assumption of a subordinate position. Some might, but others will not. As we grow used to monitoring how the patterns are flowing, we gain a valuable method of analysis.

Power and Giving or Taking Orders

We have viewed the power structure as involving what is happening between superiors and subordinates. Now look at that communication known as an *order*. We see it as a verbalization that specifies what the superior wants the subordinate to do. For example, all of the following are orders:

- Get those clothes up from the floor right now, Jenny.
- I would suggest that you read Chapter Seven for the quiz on Monday.
- Be ready at six.
- I need the analysis on my desk by lunchtime.
- Hey, kid. Move your tail unloading those boxes right now or pick up your time.

Each statement expresses what the superior wants the subordinate to do, but not all the orders are equally effective. To be called "kid" or to be told to pick up clothes "right now" might get the action done, but both partners in the dyad pay the price of ill feelings. We find that price too high.

Orders are a normal part of each day in social and business settings. Personal dyads suffer if those orders undermine the ongoing relationship. When superiors cannot communicate the actions they want done without engendering hostility, business dyads deteriorate, causing low morale and costly turnover. What do you think that child picking up her clothes is muttering to herself? What do you imagine "Kid" is thinking when he unloads the boxes. Surely it is nothing that would bind together an ongoing dyadic relationship.

We are not saying that anger or venting irritation is necessarily bad. We have a right to express how we feel, and other people should know the effects of something they have done. The price of that expression, however, might be receiver negativism. This problem surfaces when most of the orders, regardless of feeling or context, form a predictable pattern. When that pattern is usually anger or confusion or demoralization, the communication needs work.

- If dyads involve power

  and
- power involves superior and subordinate positions

  and

- superiors express desired actions in orders,

then

- we all need to give effective orders.

Whether participants are on the giving or receiving end of the communication, five principles for structuring orders should help the analysis of what is happening. An order will be more likely to effect action without harming the dyad if its formulator constructs it so that it shows the five qualities that follow.

1. Sensitivity to the Recognized Goal Goals are nebulous things. Some are short-term, and some are to be accomplished at a distant time. When a pressing task has to be done *now*, people easily forget that other goals exist too. Screaming at children or humiliating a worker might get the task done now, but future goals are placed in jeopardy. Certainly parents want the child's room tidied, but they also want the child to learn responsibility for possessions and respect for the parents. Certainly a dock supervisor wants the boxes unloaded, but that same supervisor also wants experienced workers to unload boxes the next day *and* the next week with some semblance of pleasantness.

Satisfaction of an immediate goal might be at the cost of a distant one. Supervisors must keep such a price in mind. The job can get done; but, if the entire staff quits as a result, what has been gained? Accomplishing the immediate goal while solidifying future goals requires delicate interactional skills.

2. Sensitivity to the Other Person Like all messages, orders are part of a process. Remember the receiver at the other end of the transmission. The two of you are in this process together. You can do little to improve the translation of the message if the receiver sets up entropy through defensiveness. Little that the receiver does will improve the transmission if entropy is too strong on the sender's part. Each member of the dyad is responsible for the communication exchange. Rather than blame the other, both superior and subordinate can keep the needs, shortcomings, and skills of each other firmly in mind while transacting order messages. The custom crafting of such communications enhances minimum entropy decoding.

3. Sensitivity to the Communication Options Available There are literally limitless ways to issue an order. All the symbol systems discussed in Chapter 3 stand ready as the raw materials from which a message can be constructed. The successful superior monitors and alters those symbols based on analysis of feedback received. Costs in human and material loss have to be weighed carefully.

A recent magazine article describes the 1974 negotiations for a Chief Executive Officer of Revlon.[11] The firm's founder, Charles Revson, ran his company in a management style, "from brutal to savage." Vice presidents left so often that there seemed to be a "revolving door" to the executive suite. Revson believed that "Creative people are like a wet towel. You wring them out and pick up another one." Nearing retirement, Revson knew he needed a successor and chose Michel Bergerac, then president of European operations at ITT. Bergerac knew about Revson's communication patterns and, protecting himself, demanded a five-year contract that would guarantee salary whether he continued to work or not. The price of Revson's way of communicating with his subordinates was $4.5 million. Setting aside for a moment the ethics of "wringing people out," it seems to us that Revson used his options badly.

4. Sensitivity to What the Communication Does to the Order Giver Unless one has a company that is as solid as Revlon *and* one can run it at will, no superior can forget the price paid for poor order giving. Misunderstood communications result in costly errors, for it is hard to decode insensitive orders correctly when workers are hostile. Superiors have the responsibility for certain output. Slipped schedules or missed deadlines earn little sympathy from corporate administrators. A superior has to learn what such lack of timeliness indicates. It might be that, purposefully or accidentally, orders are being decoded incorrectly. In either case, the superior has a problem.

Beside the damage to a superior's career, poor order giving results in a lot of hard feelings. The psychic energy a superior expends fuming about a nasty encounter with a subordinate could be used far more profitably in other ways. The price might not be $4.5 million, but we still find it too high.

5. Sensitivity to Clarity No order can be fulfilled if its message is garbled. In the final analysis, all the communication principles discussed in Part I of this book come into play. Articulating the message, looking for feedback, and checking for active listening must be an inherent part of the process. Whether the entropy originates in poor selection of symbols ("Which spreadsheet does he want?"), sloppy speaking ("Did she say, 'Keep the disk' or 'Clear the desk'?"), or mistaking hearing for listening ("Yeah, yeah, I heard you."), the entropy will throw the message into disarray unless constant vigil is kept. When an order is given so that its sender recognizes the levels of goals, takes note of the other person in the dyad, selects the symbols carefully, remembers that orders reflect on the order-giver too, and practices good communication, that order is competently given. When subordinates monitor orders through these considerations, it is likely that they will be able to fulfill them as the superior intended.

WRAP-UP

Dyads form the basis of most communication. Messages are decoded by a single receiver, no matter how many of these receivers there might be. When there are actually just two persons present, this dyad forms a special communication setting well worth studying.

In business settings, interviews begin careers. They must be handled carefully through preparation, recognition of questions, and enactment of expected courtesies.

Interpersonal relations evolve from the daily communications between professionals in a dyad. Successful communications lead to productive networks. Badly handled communications result in destructive conflict. Networks and conflict must be managed carefully. Both can be useful and a source of growth.

Power structures are built around the concepts of superior and subordinate. Dyads are concerned with power, especially—but not exclusively—in the business setting. Understanding superior and subordinate responsibilities in power dyads require analyzing several areas of concern.

Finally, power structures eventually involve the giving and taking of orders. The competent superior must know how to structure this important communication. The dyadic partner must know how to monitor it. Together, the superior and subordinate enter into a process of giving and taking orders to implement actions. The business day is largely composed of such interaction.

ACTIVITIES

1. Prepare a simulated interview in which one person serves as an interviewer and another person acts as a candidate for some

professional (and technical) position. The class will observe this simulation and then analyze:

   a. What kinds of questions, open or closed, were used at what point in the interview?
   b. What did the nonverbals of both participants tell about what was happening?
   c. What information seemed to dominate the interview?
   d. Were there any answers to questions that damaged the credibility of either person? How did they do their damage?
   e. How would this candidate compare with other (imagined) candidates for such a position?

2. Select any *five* dyads in which you participated over the last week. Using the definition of power discussed in this chapter, how did power shift during those dyads? How do you account for the shift?
3. Divide the class into dyads. One member of the dyad assumes the role of employer, and the other assumes the role of subordinate. Each dyad should present to the class two brief presentations:
   a. One in which a set of orders is well given.
   b. One in which a set of orders is given poorly.
4. Recall a dyad during which you were in a subordinate role, and the superior gave such poor orders you were not able to carry them out. Using the five guidelines for good order giving suggested in this chapter, describe to the class how your superior could have improved the orders given to you.
5. Locate any network to which you belong (or would like to belong). What kinds of dyads would help you maintain or gain acceptance into that network? How do you establish such dyads?

NOTES

1. Rollo May, *Love and Will* (New York: Dell, 1969), 307.
2. John Stewart, ed., *Bridges Not Walls*, 3rd ed. (Reading, Mass.: Addison-Wesley, 1982), 6.
3. Cal W. Downs, G. Paul Smeyak, and Ernest Martin, *Professional Interviewing* (New York: Harper & Row, 1980), 5–6.
4. For a comparison of how the written poll takers such as George Gallup use such open and closed questions to secure information, see Stanley L. Payne, *The Art of Asking Questions* (Princeton: Princeton Univ. Press, 1980.)

5. Alan Garner, *Conversationally Speaking* (New York: McGraw-Hill, 1980), 6.

6. Downs, Smeyak, and Martin, *Professional,* 309. Although the authors cite this technique in the broadcast interview, the dangers are equally applicable to the job candidate.

7. Rudolph F. Verderber and Kathleen S. Verderber, *Inter-Act,* 3rd ed. (Belmont, Calif.: Wadsworth, 1983), 188. Although we have structured the types of conflict differently from Verderber and Verderber, much of our treatment of conflict has been influenced by their work.

8. Verderber and Verderber, *Inter-Act,* 191–95.

9. Andrew J. DuBrin, *Foundations of Organizational Behavior* (Englewood Cliffs, N.J.: Prentice-Hall, 1984), 387–88. He cites David Kipnis, Stuart M. Schmidt, and Ian Wilkinson, "Intraorganizational Influence Tactics: Explorations in Getting One's Way," *Journal of Applied Psychology* (August 1980), 440–52.

10. DuBrin, *Foundations,* 391.

11. Patricia O'Toole, "Corporate Messiah," *Savvy,* (May 1984), 57–62.

CHAPTER ELEVEN

SMALL-GROUP THEORY

In a journal article written for electrical engineering students, Paul Rega tells the story of a ceramic engineer who tried to describe his new idea in a series of dyadic communications.[1] First the research supervisor and then the marketing manager listened to his description of a type of glass that turned dark when exposed to sunlight and turned clear again when brought into a normally lighted room. Both thought the idea "interesting," and they placed it on the agenda for the marketing meeting. The engineer met with the marketing group, elaborated on the technical formula, described the nature of the crystal properties at length, and ended by asking, "What should I do with this piece of glass?" The group responded with silence and then the senior vice president answered the question by making a graphically obscene gesture. The meeting members convulsed with laughter.

The engineer was crushed, so much so that he took his glass, made up toys with it, and gave these to children. He tried for two years to forget the humiliation of that meeting. Fortunately, a friend in the optical division remembered the glass, and eventually it became the basis of an entire eyeglass industry. The invention, of course, was the light-sensitive glass, used in lenses that can double as sunglasses.

Some of us experience a strange ambivalence about working in groups. We suspect that many good ideas, like the one the ceramic engineer proposed, are lost in meetings. On one hand, we visualize all the time spent in chatter with a committee, and we moan, "I could have done twice the work in half the time myself." On the other hand, however, the thought of having no part in decisions that affect our lives strikes us as unacceptable. And, in fact, groups are integral to the modern decision-making process. We may not always like working in them; we may even conclude that they lose many good ideas; but we know they cannot be ignored. Most important, group settings provide an excellent opportunity for learning, social interaction, and problem solving.

This chapter presents some basic group theory. We begin by suggesting a definition of group as a distinct communication setting. After considering the types of groups, their advantages and disadvantages, we focus on one: the problem-solving task group that forms the basis of most business meetings. Finally, we shall examine a few of the special adaptations to group process that modern industries have made. We do not like to think that the great successor to the photosensitive glass idea will someday be lost in one of your meetings.

COMPONENTS OF GROUPS

Would you call any of the following a *group*?

- Seven persons standing at an intersection waiting for a green light.
- Thirty thousand fans gathered on a Saturday afternoon at the football stadium.
- Twenty-five students sitting in class on the first day of a new semester.

None of these falls within our definition of a group, although all are collections of people. We define a *group* as a type of communication pattern in which three or more people assemble in person for a special purpose, so that each participant has a chance for equal awareness of and interaction with every other person. During the interaction, norms and a sense of belonging develop as a means of reaching consensus. Now, we must separate that mouthful into its constituent parts.

Numbers

No magic number suddenly determines that a collection of people can be called a group, but we propose that certain communication factors should be considered. We have categorized dyads as a special sort of pattern. In Chapter 10, we examined the one-on-one encounter as defined by its own component parts. Keeping to that pattern, we assert that a group has to be more than two people.

No upper number exists, except that each person must be able to interact with and have a reaction to every other person. The football crowd of thirty thousand is disqualified on this constraint.

Some claim that the upper limit of small-group size is as low as five members, because numbers larger than five change the communication pattern.[2] Others say that large groups actually increase productivity up to a point but that, for decision-making bodies, optimum size is five or six.[3] Still others note that three must be the lower number, but that circumstances determine the upper limit: "In most groups the upper limit will be about fifteen, in some groups it may only be seven, while in still others it may exceed twenty."[4]

Putting all these statements together, we say that a small problem-solving group should number at least three to five and no more than ten people. This seems few enough for each person to interact directly with each of the others. For practical reasons, we also prefer an odd number.

Purpose

A second necessary part to our definition is the requirement that the persons all agree that they are using each other to achieve a goal. This goal could be solving a problem, sharing information, or just interacting socially. Note that, in the three examples at the beginning of the chapter, the people might be aware of one another, but they have no plan for interaction. Look at the seven people at that intersection. They did not come together in order to engage in a particular communication pattern for a special purpose.

Norms

Perhaps because of the awareness group members have of each other, they develop informal rules of procedure called *norms*. Group norms evolve out of the need for structure. Some groups regulate themselves so that members may speak only after securing permission from an acknowledged leader. Others speak when members wish, cutting each other off. Certain topics might be inappropriate in one group, while anything goes in another. Whenever people gather in true group membership, they tend to develop these regulatory patterns. Random associations of people do not.

Cohesion

As the term *cohesion* implies, members develop feelings that they belong, they stick together, they adhere. We are fascinated by how quickly a group begins to speak of themselves as "we" and everyone else as "they."

Industry provides a natural environment for cohesion by virtue of competition between companies and competition between groups in the same company. Even in the temporary gathering of a classroom group assignment, a sense of belonging develops that separates the members from the others in class. B. Aubrey Fisher calls this quality "groupness," and claims that it is possible to detect growing levels of cohesion.[5]

Consensus

The final part of our definition concerns the product of groups: decisional consensus. The term does not necessarily mean agreement. It does mean that each person in the group, although perhaps not completely liking the decision, *can accept and live with it.* A consensus is much like a bargained price. If you want to sell your guitar for two hundred dollars, and your roommate offers you one hundred for it, you will probably bargain. Suppose you settle on one hundred and fifty as a fair price. Think about that process: you are getting fifty dollars less than you want, and your roommate is paying fifty dollars more. Neither of you likes that compromise completely; but, if you both feel that the price is fair, you have reached consensus.

Special Collections of People

Look again at the three examples that began this discussion. Suppose that one person among those at the intersection suddenly fell to the sidewalk, gasping for breath. It is possible that the other six might begin to function as a group. They begin to communicate in terms of a specific purpose: helping the stricken person. They become aware of each other and develop norms for activity: "You go telephone for an ambulance"; "Don't touch him. She knows CPR; let her do it"; "Cover him with your coat." They arrive at a consensus of what to do by using each other and, along the way, identify themselves as part of the group. "Yeah, officer. We were the six standing here when he fell down. Those others came later."

In the same way, the football fans might separate into small enough numbers to develop communications in true group function. A classroom of students eventually might also. As with many of the concepts discussed in this book, whether a collection of people meets our definition depends on the communication pattern. Not every association of persons is a group; and the only way to tell is by the nature of their communication.

TYPES OF GROUPS

With a definition of group in place, we can now consider some of the various types that exist. We will not discuss many group formats: a case conference, a therapy session, and role-playing encounters, for example. These are important, but we have decided to present briefly those types that are the most common and then to move on to the one most used in the business and professional world.

Brainstorming

The non-evaluative means of generating ideas called brainstorming is quite productive. Basically, a group of people agrees to list ideas about some topic without evaluating any of them during the brainstorming session. The lack of evaluation is important, since no idea, regardless of how off-the-wall it may seem, should be eliminated at this point.

The ceramic engineer with the peculiar glass that changed color could have used a brainstorming session. Rather than dismissing his "What can I do with my glass?" question, the participants could have listed a number of ideas. If the atmosphere of the session is open, many suggestions might be generated. A recorder should be appointed to write down the ideas. No comment or evaluation ought to take place. When the list is complete, it should be duplicated and distributed to those who generated it in the first place. At a later time, after they have had a chance to think, the group members or another group meet again to examine each suggestion for its worth.

Spontaneous Meetings

Many social or semisocial encounters evolve into groups. The people with whom you eat lunch three times a week probably function as a group. Between the sandwich and the apple, you all advise Jim how to handle his superior, Liz how to get the best bank loan for her car, and Carlo how to bounce back from his broken engagement.

Three or four people stopping by your office might begin to speculate about better ways to market the product you have developed. These chance meetings, often informal in nature, are extremely important to the social and professional life we lead.

Panels and Symposia

Sometime before most of you left high school, you probably participated in something called a panel. It may have been in a social science class where six of you prepared formal oral reports around some central topic such as "The Emerging Third World Nations of the African Gold Coast." Actually, what you participated in is a symposium.

A *symposium* is a discussion among experts of some central topic. Each expert speaks in turn, without interruption, on one portion of the topic. The symposium is nearly always planned in advance, at least to the point of dividing up the topic among the participants, and is usually presented for the benefit of an observing audience.

A *panel*, on the other hand, is also a discussion among experts for the benefit of an audience, but each expert is free to enter and leave the discussion at any time. Participants do not give planned speeches, as in the symposium, but rather interact with all of the other participants randomly. The audience "overhears" their comments.

Forum

The term *forum* has come to have several meanings, but each of them is related to its core distinction. A forum is an open opportunity for anybody to speak who wishes to do so. Thus, a town meeting is sometimes called a forum, as is the question-and-answer period that often follows a guest speaker's lecture. Basically, the idea of forum is to give those who might not be considered experts a chance to have their say. Whenever an audience is invited to comment, question, or object, such a portion of the communication is called a forum.

Task Meetings

In this chapter we will focus on the task meeting and its cousin, the agenda problem-solving group. Our decision is purely pragmatic: the business world uses this type of discussion format frequently, and a professional must learn how to function well within it.

A task meeting is one where a group (in the full sense of our definition) assembles in order to perform some specified task. The task might be highly complex—critiquing the new product before commitment to production—or relatively simple—suggesting an effective way to reduce litter in the hallways. The important feature of this type of meeting is that the participants know that a specific problem must be solved, and they assemble to accomplish that task.

In the sense we use the word *task,* we assume that there are two kinds of task groups: those with a minimum of outside structure imposed on them and those where a particular structure is required. In the minimum-structure group, often the only instruction given is the deadline by which a solution must be found and perhaps implemented. How participants are to reach that solution is up to them. The structured group is quite another story.

Agenda Problem-Solving Meetings

The agenda meeting is a task-group meeting with an imposed structure. An *agenda* is a set of actions the group is expected to follow, listed in the order the group should consider them. Thus, the path the group takes to arrive at the solution to its assigned problem has been determined externally.

An agenda might be prepared by the meeting chairperson, by a subcommittee, or even by a person (such as the corporate president) outside the group membership. This order of business is usually distributed well in advance of the meeting so that participants have a chance to reflect on what is going to happen when.

Agendas are common organizational devices, especially when the meeting is to take place between companies. If several corporations are cooperating on a given effort, their meetings usually concern a problem that needs to be addressed by all of them. In this instance, developing an agenda ensures that each participant will prepare fully before the actual time of the encounter.[6]

DISADVANTAGES AND ADVANTAGES OF GROUP PROCESS

Most of us realize that when we work alone on a problem, the process is different from when a single friend joins us. We also know that if four other friends come along to join, the combination of six alters what happens in a way more complex than simply adding four to two. Consider the disadvantages and advantages group activity entails.

Disadvantages

Certain cost factors must be considered when people work in groups. These costs are not usually financial, although they might be; but they do involve a price that participants often must pay.

Inefficiency Groups bring together people; and, as social creatures, these people will interact. They socialize, ramble, argue, become distracted, and—in many other negative ways—tend to produce with great effort but little efficiency.

Time Consumption Related to the inefficiency of groups is the reality of how much time they take. Group decisions are not speedy. When the vice president wants a decision by 5:00 P.M. Friday and it is now 3:00 P.M., do not convene a group. Six persons meeting to solve a problem will probably take much longer to do it than a single person would.

Entropic Communication Losses Remember that for each separate communication and for each separate receiver, you have multiple loops to complete (see the communication model, Figure 2-11). Lurking at each point is entropy potential. The more people there are, the more the potential rises. Listening problems, misunderstandings, partial decoding, and sender-receiver biases are to be expected.

Submersion of Minority Opinion It is difficult to stand alone when everyone else in the crowd agrees to something. How well most of us know that! Demos and O'Neill, clinical psychologists, trace the use of "groupthink" in reformation of rebellious prisoners in Communist China. Individual members in groups found it virtually impossible to stand against the ideologics that the rest of the group held.[7] Even in a less coercive setting, the holder of a minority opinion might decide to remain silent or to verbally agree with the views of the group.

Pressure to Conform Related to the submersion of minority feelings might be the direct pressure of members on the one who holds a different view. Participants in groups attempt to change the minds of the others through legitimate persuasion. When that fails, some use more coercive strategy: "Come on, Jack. We don't want to be here all night. Sign this report so we can have a unanimous decision, and we can all go home." Jack might sign, not because he agrees, but because he feels pressured to do so.

Differing Levels of Motivation Not everyone comes together for the same reason. In any group, some people might believe strongly in a position, some might know very little about what is discussed, some might come because they have to, and some might come to impress the supervisor. Each enters into the process differently and each, unfortunately, affects all of the others.

Different Levels of Ability Even if all participants have a strong commitment to solving the problem in the best way they can, potential problems exist. Some people have analytical abilities so far beyond the skills of the others that they become bored or overbearing. Some have verbal skills so eloquent that they sway the rest. Some are so deficient in a particular background skill that they cannot follow the train of thought. Yet, each of these is a member of the group, and the group has to adjust itself to its people.

Unacceptable Group Norms Groups sometimes adopt ways of proceeding that are irritating or abhorrent to some of the members. If Sarah has to ask permission of the group leader to speak in a given group and such a procedure bothers her, she will have a difficult time. Clashing of norms, differing values on how to proceed, or rules adopted by the others that seem silly will be the price of staying with some groups.

Advantages

Despite their costs, the widespread use of groups to solve problems, gather information, and provide member support tells us that some significant benefits are possible.

Expectation of Participatory Rights The Western world has established and accepted the idea that people have a right to speak on matters that concern them. Such orientation makes it difficult to obey without feeling that we have contributed to the decision making. Group structure is a setting where we can express our views in hope of affecting the final outcome. Most of us find that we expect to be included in discussions of things that will affect us.

Quality of Decisions Increases Research indicates that the type of decision reached in group settings might be of better quality than that reached unilaterally.[8] Quality is not a magic result of the increased numbers; rather, the process of subjecting ideas to the scrutiny of others tends to improve thinking and reveal flaws.

Possibility for Added Information By simple arithmetic, we know that many people bring many more experiences to a problem than a single person does. These experiences add to the idea pool. Data known to a single member might change the way an entire group thinks; if that person is not within the group, the data remain unknown. From the expanded pool of ideas, solutions emerge that are beyond the capabilities of any single member.

Possibility for Expanded Abilities Personal differences in abilities and interests can be a disadvantage, but they can also be a distinct advantage. By having one or more members who know how to do something the others do not, an entire group's potential for problem solving is expanded. We remember when a statistician in a group saved the other members from a serious mistake in one piece of evidence. He pointed out the irregularity of the way the numbers were applied. The conclusion drawn from the evidence turned out to be incorrect, but the rest of the group would not have known that if the statistician had not been there.

Enhanced Satisfaction Related to our orientation for voicing our opinions seems to be a tendency for us to more easily accept a solution if we have participated in its formulation than if it has been handed to us. Managers should especially be aware of this tendency. When the administration sends down instructions for a directive, especially if it involves a controversial issue, allow representatives of the affected groups to participate in the formulation of the directive. Even difficult decisions, such as speeded-up work schedules or selective layoffs, appear more palatable if those who bear the brunt of the decision participate in it.[9]

Satisfaction of Social Needs Remember that people are social creatures. A group experience can satisfy the need for acceptance and even self-esteem. Humans are not loners. Working with others is a natural part of our makeup. It may be that the presence of other people serves as a catalyst for people to strive more than they would if they worked alone.

Enhancement of Communicative Skills The group setting permits some people to express themselves more freely than they would in dyads or in a large public gathering. Groups become places to practice competent communication and to gain the reputation for doing so. Not everyone is equally effective in every communication. Stories abound about Presidents Lyndon Johnson and Richard Nixon, who could be extremely persuasive and personable in the small-group setting, but who never quite mastered the large public presentation. Groups might be your particular place to shine. Not only can you learn to expand your skills there, but you can also gain a reputation for being a good group member.

TYPES OF DISCUSSION QUESTIONS

Now that we have presented a definition of groups and discussed the various kinds and their disadvantages and advantages, we move on to consider what groups actually discuss. Unfortunately, too much group time and effort is often spent deciding just what the topic for discussion is supposed to be. Wording and consideration of the general topic are extremely important. We present some guidelines for clear topic statements, and then consider what the three different types of topics are.

As we noted in Chapter 3, the entropy potential of language is enormous. To forestall some of the problems language use can cause in group discussion, consider the following as minimum standards for any topic under discussion.

- A topic should be stated as a *question.*
- The wording should be as clear as possible.
- Any biased words that would tend to slant the discussion should be eliminated.
- The topic should be researchable to those who will participate in the discussion.
- The topic should be of interest to the group members.

Whenever group members say "Why are we meeting today?", someone has been extremely derelict. A group should know specifically what question will be considered; that question should have been stated so clearly that nobody arrives prepared to discuss something other than what was intended; that question should have been examined for any prejudicial words that would unnecessarily limit possible answers; that question should also be within the province of information available and interest to the members. If these guidelines are observed *before* the first word is spoken at the actual meeting, the chance of productivity increases.

The types of questions that groups discuss have traditionally been divided into three sorts: questions of fact, value, and policy. Yes, two of these are the same terms we have considered in other contexts. Facts and values are so much a part of the human experience that they continue to surface in various communications.

Questions of Fact

When we considered evidence in Chapter 8, we defined *fact* as some event that exists as our verification says it does. Fact statements make a claim for the accuracy, the existence, and the verifiability of some event. The discussion of factual questions operates similarly, by proposing a question about something that can be measured in some way. In discussing that measurable something, a group assumes that both existence and accuracy of observation are present.

Some fact discussion questions follow:

- Has organized crime infiltrated Peaceful Valley's township government?
- Does zinc or paint provide longer durability as a coating for the major container?
- Did the Easter Island people practice a form of advanced science in the sixth century that allowed them to build and move heavy statues?
- Is Stonehenge an astronomical measuring system?

You might have strong opinions about any of these questions, but notice that the discussion is not about those opinions. It concerns the observable and verifiable data surrounding something outside you.

Questions of Value

Values are part of an individual, and in Chapter 5 we linked them to the Versions of Me model (Figure 5-1). Values, thus, are statements about your beliefs, opinions, and inner activities. Discussion questions about values presuppose that group members will talk about what they and others feel and believe about some event.

Value questions for discussion might be:

- Is capital punishment ever justified?

- Is it right that Mega Corporation should enter the military weapons market?
- Is it legitimate to withhold information from the customer?
- What's wrong with young people these days?

Many facts are probably associated with each of these questions, and they might be advanced as reasons for holding a particular opinion. The bottom line, however, is that each of them asks how the participants think and feel about these events. Value questions have a "to me" quality about them. They are important for understanding what you and others believe.

Questions of Policy

Questions of policy concern what should or will be *done* regarding something. Do not be misled by the auxiliary verbs should and will. We do not use them to designate value or personal feelings about an action, but as neutrals to designate action. When you say "We should paint this room green," you mean that you advocate changing the color of the room to green. No implication is made that you like to paint or prefer green.

The business community is always in the midst of change. Products are developed to take the place of existing ones; theories are explored because they improve designs or reduce costs; methodologies are implemented when they are found superior to existing procedures. The heart of business is policy, and so discussions about what is to be done abound.

Some policy discussion questions are:

- Should we change our supplier from Automax to Quality Products?
- Should Ralph go to the Institute Conference in Paris?
- Should we authorize a 10-hour day for our production facility?
- Should the United States Presidency be restricted to a single six-year term?

Many values and facts will be expressed in each policy discussion. People will speak about what they believe and what the data show, but the purpose of the discussion is to decide to *do* something that is not now done.

Discussions usually involve all three types of questions. Values cannot be separated from the people who hold them. Neither can the human desire for factual bases. When values and facts are expressed, policy changes often follow. The importance in knowing that all three types exist helps you keep the goal of a particular meeting in mind. If the question is one of value, your answer does not require doing something about a given problem, but rather increasing your understanding of others' viewpoints concerning it. If the question is one of fact, your answer will increase information about the issues. When the question concerns policy, however, your answer will advocate action, usually changing one thing into something else.

CORPORATE FORMS OF GROUP MEETINGS

Business organizations use methods that work. If an idea has merit but needs change, they amend it. What sometimes emerges is a unique set of practices, tailored to the specific activities of the business. We have observed a number of businesses, in both the United States and Europe, and have concluded that group meetings are such custom-tailored activities. We cannot claim that all businesses function as do those we have observed, but we do think it is worth your while to consider how the general ideas developed in this chapter seem to be applied. Furthermore, because one of us works within a large technical industry and has attended each of the following types of corporate meetings, some insights at the participant level are possible.

We examine first in-house meetings and then intercompany meetings. The names given to some of the meetings reflect the specific terminology used by General Electric, although we believe that parallels exist throughout other corporations.

In-House Meetings

The appointment calendar of a mid- or upper-level manager in a large corporation can be extremely revealing. When the manager is a technical professional, the word *meeting* tends to appear almost daily. Even when those meetings in dyads are eliminated, what remains is a picture of many hours spent each week in group problem solving.

Many group meetings are *in-house.* As used by industry, the term means meetings of people in the same company. Where the meeting takes place is irrelevant to the label's use. An in-house meeting might happen at a conference center, the corporate offices halfway across the country, or just down the hall from where the group members work daily. The important variable is that its participants are part of the same corporation. The variety of in-house meetings is reflected in the six we will now discuss.

Regularly Scheduled Meetings We did not observe too many "We're meeting because it's Wednesday" groups in American businesses. In Germany, however, our sources at both a large automotive corporation and a mining research corporation reported widespread regular meetings in German businesses. Their own companies held such meetings, we were assured, at each worker level. The automotive representative claimed that his subordinates meet every morning without fail at 7:00 A.M. to preview the workday.[10] When people gather *only* because they have always met at a given time, problems can be anticipated; but, then too, the flow of information is kept up to date. We were assured that such meetings are not at all unusual in Germany.

In the United States, we discovered, upper management often does use a regularly scheduled monthly meeting for reviewing finance. The assumption is that a major assessment of the corporate profit picture is needed at least every thirty days. These meetings are always agenda meetings, attended by personnel at department-head level and above. Other employees such as accountants, resource people, or specialists are invited only as needed.[11]

Crisis Meetings In a business, a crisis is any problem of such magnitude that a project or major corporate policy is in jeopardy. In Chapter 6, we spoke at length about a company's concern for acceptance of contract. As you review those steps, notice how easily problems in a contracted project might arise. Some of these might necessitate crisis meetings. Usually such meetings concern the following topics:

- Unanticipated trouble with some aspect of the project.
- Slipped schedule due to in-house or supplier problem.
- Unanticipated results from some testing.
- Customer desire for change of contract scope.

A crisis meeting, by its very nature, seldom allows much notice to be given to its participants. No more than twenty-four hours usually pass between the first recognition of the crisis and the meeting called to deal with it. Agenda preparation may or may not be used, but specific information is expected both by those who call the meeting and those who attend.

Orientation Meetings These meetings are usually semisocial in nature, at least to the degree that they have a "light" tone to them. They are called to introduce a new manager to the group, for example, or to discuss some new corporate policy. Both of these are serious changes, and workers need to feel informed about them; but they are most usually done in an upbeat manner. The manager will say a few words of encouragement, insisting that "We're going to do great things together." A new policy is explained and, in both cases, people are expected to comment and ask questions. The tone is often quite festive.

Not quite so festive, but still decidedly upbeat, are proposal orientations. When a proposal for some contract is to be written or when the contract is won and work is to begin, participants in the effort gather for orientation. They are informed about the scope of the project, and they are expected to question and comment. Since each of them worked on or will work on the effort, levels of good information are presumed for all participants.

Progress Meetings The most common type of meeting for technical corporations is probably the progress meeting. Many checks and balances are built into the process of completing a contract. Because technology is so specialized, many different groups will be working on separate parts of the overall contract. The groups have to be brought together from time to time to see how their respective assignments are progressing. The progress meeting is the device used.

There are several levels of progress meetings, depending on how a particular corporation divides its staff. The following is a fairly standard division, and you should be able to see the pattern possible for different divisions.

Progress Analysis and Review (PAR) is a meeting designed for upper management. It is nontechnical in the sense that no technical problems are discussed directly. Its purpose is to bring together the upper-management personnel for a business analysis of the project (contract) in process. PAR meetings might be held frequently during the duration of a contract, and they might become crisis meetings if some unexpected business problem surfaces. They are always information-laden and carefully prepared, since direct data is required to verify a particular business conclusion. Subordinates and support staff are only invited as resources for some needed information.

Task leader meetings are the most pervasive. Here the actual working problems of design, purchasing, support systems, and experimentation are reviewed. These meetings are ordinarily attended only by those in charge of the various component groups. Personnel from those groups are often invited as resources. Thus the task leader of the controls group might invite two members of the control staff to present the problems that have surfaced in the group's work.

Those staff specialists might consume most of the time set aside for a given meeting. They come prepared with visual aids, usually transparencies, but often with charts and actual objects. They may speak for extended periods of time, explaining why their group finds itself confronting some problem. In such presentations, however, expectations are different from those of the usual public speech. The speakers *expect* to be interrupted often with questions and comments. The audience, at the same time, *expects* to enter into the prepared presentations as the need arises. Most tellingly, the speakers use their presentations in order to involve the other members in collective problem solving. In many ways, this corporate format is a *hybrid public speech.* We shall have more to say about it in Chapter 13.

Progress meetings are based on the assumption that each person present has a shared responsibility to evolve a common solution to any particular working group's difficulty. The task leaders can share information about what their staffs are doing and also elicit ideas for error detection and flaws.

Staff meetings form yet another kind of progress review. A staff is usually an entire pool of people involved with a particular function. Thus all the electrical-design personnel form a staff, as do all the purchasing members and all the legal representatives. An individual staff member might be working on any number of separate projects, each having little to do with the project of a staff colleague except that it requires a particular skill.

The manager of the drafting staff, for example, might attend ten different task leader meetings corresponding to ten different projects the drafting staff supports. During those task meetings, the staff manager might report that drawings are behind schedule in six out of the ten projects. The task leaders suggest actions, and the drafting manager probably feels the need to call a staff meeting.

Staff meetings can be extremely technical. They are called to disseminate administrative information or to address some problematic issue likely to affect the entire staff. The members of a staff group use each other to reach consensus about a policy question:

How can we improve the output of the entire drafting staff so that no project has a slipped schedule?

Unit meetings are the narrowest and most specialized of all the progress reviews. Within a staff, say the drafting group discussed above, four persons may be working on the AEGIS project. These four form the unit of drafting personnel, and it is usual to appoint a leader among them. This lead drafting person might call a unit meeting of the three others working on AEGIS. Those who attend are directly involved with the task, so attempts can be made to define what was done, where the problem surfaced, and possible approaches to solution. The unit leader also identifies limitations in available resources, bottlenecks, or need for outside expertise. These ideas are communicated to the staff manager and, finally, to the project manager. Discussion questions for this unit, like those of the staff meeting, involve policy: How can the four of us deliver all drawings for AEGIS on time?

With at least four levels of meetings going on at different times, all dealing with some aspects of a project common to all of them, the opportunity for entropy is high. Recognizing the need to improve communication, businesses have used a creative method of feedback called the *control room*. The control room is a physical place, set up with elaborate visuals that communicate at a glance what every component of a particular project might be.

If the AEGIS project, for example, is a large one, its control room will be correspondingly large. Often a conference table or some sort of meeting facility will be housed within the room. Most important, though, are the visual aids that line the walls. A person entering the room can see at a glance the names of all the different support groups working on the project. Names of task leaders, staff managers, and unit leaders are posted. A coding system is usually developed that uses colors and other symbols to communicate important information. In this way, a viewer can see:

- What deadline each support group has.
- How close the work of the group is to completion.
- What group is behind deadline.
- What the projected new deadline for completion is.

Figure 11-1 sketches one version of such a room. A stranger might find little that is meaningful in all the colored dots and odd-

Figure 11-1 Photograph of a typical control room. (Courtesy of General Electric Co.)

shaped flags. The control room is meant to communicate the status of a given project only to those who have been taught its symbol system. Task leaders or any individual worker may check several times a day to see the changing status of a given component. Personnel stop into the control room often, especially if some support system is having problems. Meetings are held in the room so that the visuals can save review time. In all, the control room serves as a communication aid, helping important participants keep track of what happens. It is one means of entropy control that industry uses to good effect.

Skip-Level Meetings In a large corporation, upper management is easily insulated from open communication with subordinates. As we suggested when discussing the progress review, the trend is to meet with people at a similar responsibility level. These meetings may be efficient, but important feelings and ideas remain unknown. The upper manager has a buffer of several levels and is shielded from what is happening in the trenches. Conversely, a member of a given unit might never discuss an innovation with a task leader but would have to count on the unit leader's sending the information up to the staff manager who, in turn, would have to present it to the task leader. If this unit member had a grievance directed at the staff manager, a grievance perhaps shared by others in the unit, it might be difficult to express that grievance without being blocked by the manager. Especially in corporations where unions do not exist or channels for grievance are ambiguous, insulation of upper-level managers is a problem. The skip-level meeting is industry's way of dealing with it.

A skip-level meeting brings together a group with the supervisor of the group's supervisor. Several levels might even be skipped. For example, a vice president might sit with the production staff in order to hear their concerns and ideas. None of the intervening supervisors is present at these meetings. The idea is to remove the usual supervisors so that their subordinates feel free to express whatever they wish.

Such meetings are controversial, for the technique sets aside the usual chain of command. Its benefits are thought to be worth the anxiety surrounding it:

- Open and direct communication with a supervisor at a level not usually available to the workers.
- Freedom for both levels to hear information that is usually difficult to secure.
- Identification of emerging problems that have been hidden from the upper manager.

Such meetings are not held *only* when the immediate supervisor is suspect. Entropic communication losses are expected throughout the many levels of corporation command. Even well-meaning supervisors might forget or underestimate some idea raised among their subordinates, and that idea is never passed along. The skip-level meeting attempts to reduce such loss. If the immediate supervisor has been operating poorly, of course, the workers might find the skip-level meeting the only opportunity to express their concern.

Quality Circles A recent development in corporations is the systematic use of workers at all levels to solve problems of corporate concern. Sometimes called the quality control circle (QCC) or participatory circles, these groups gather people from across the lines of command and skill. Everyone from the lowest-paid production-line apprentice to the chief executive officer is assumed to have some ideas for solutions to common problems.

The quality circle is attributed to the work of W. Edwards Deming, an American statistician who taught quality-control methods after World War II. The United States was slow to accept his ideas, and he began working with the Allied-occupation government in Japan. The Japanese saw the promise of what he taught, and he is usually acknowledged as a major contributor to Japanese achievements in quality products. Yearly, Japanese industries give the Deming Award to the company among them that shows the greatest gain in quality. For his work, Deming was awarded the Second Order Medal of the Sacred Treasure by Emperor Hirohito.[12]

Deming taught the Japanese to identify quality-control problems statistically, but it was an engineering professor at Tokyo University, Dr. Kaoru Ishikawa, who tied the method to participatory-worker involvement. Approximately ten million people participate regularly in quality circles in Japan, and circles are beginning throughout Asia, North and South America, Australia, Africa, and Europe.[13]

Businesses in the United States are rediscovering the advantages of quality circles, and corporations such as Boeing, Lockheed, Hughes and Northrup, Reynolds Tobacco, Uniroyal, and "hundreds of others" are implementing them.[14] This group structure is likely to become a part of the corporate world. As workers strive to control the quality of their products by determining what hampers that quality, they seem to do so effectively in quality-circle groups.

Intercompany Meetings

Frequently, more than one company is involved in a given project. Sometimes there is a peer relationship among them. A customer signs a contract for several different tasks split among several companies, with each task leading to the completion of a common overall project. Although the companies work on portions of the task independently of each other, they need to meet for an overview of the entire picture.

More often, a prime contractor (some large corporation) secures a contract for a project and then hires subcontractors to perform some of the functions. The prime contractor becomes the customer to those who have specialized skills needed on the overall project.

Whether the companies are peers or customer and supplier, the specifics of each portion of work are spelled out by contract. The contract often mandates a certain amount of meeting time because, under such arrangements, progress reviews are necessary periodically; frequent meetings at different levels are to be expected.

Cooperative Contracts As in any activity when "guests are expected," intercompany meetings are always more formal, more prepared, and more cautious than in-house meetings. Such face-to-face group meetings among those who must share information on a project are usually prefaced with many prior communications: phone calls, dyads, even smaller in-person group meetings. The participants might know each other fairly well or hardly at all. They come together to report and sometimes to defend their work to others whose job it is to provide some other skill to the overall contract. In this spirit of sharing and cooperation on a common goal, they function as a true problem-solving group.

These meetings usually utilize the hybrid public-speech model we have noted. Each company presents its progress to the overall group. One person or several speak orally for extended periods of time. Visual aids are almost always required, and these are much more elaborate than those used in-house. Presentations are prepared carefully. A "dry-run" in-house before the intercompany meeting is usual. During the presentation, the receivers are free to comment and interrupt the speaker at will. Suggestions are made at any point, and a diverse comment might send the group searching for information not anticipated by the presenter. The group's job is to share in the criticism and support of the presenter's work. That person, in turn, speaks as representative of an entire corporation. In Chapter 13, we shall make some suggestions for you who are preparing for this world of hybrid public speaking.

This meeting type almost always uses an agenda, which is sent to participants well in advance. The agenda permits preparation of a presentation that draws on public speaking traditions, yet takes place in a problem-solving group.

Customer Review The most common intercompany meeting is the customer review. It is, after all, the people purchasing the company's product who must be kept satisfied. For the elaborate and expensive service product of developmental technical corporations, that customer is often the federal government. Those the government hires can, in turn, hire subcontractors for portions of the contract they cannot complete in-house.

Regardless of the customer's identity, the prime contractor tends to call frequent review meetings. These are always carefully prepared and quite formal encounters. An agenda is almost always used, and the subcontractor readies the presentations based on the agenda.

The hybrid public-speaking model is the most usual format. Typically, the reporting subcontractor addresses the group, while the audience of the prime contractor and all other subcontractors summoned to that review range themselves around a table or conference room. Visuals are elaborate. Their absence, at least in the last ten years or so, would be considered peculiar. The persons reporting almost always have had a dry run with criticism in-house. All members of the audience are expected to comment at will. The presenter might begin with a carefully prepared sequence of comments, only to have them interrupted by a thought raised by some questioner. The ability to adapt information to the needs of the audience is paramount. Documentation and specific proof are expected. Their absence or a presenter's lack of credibility is seen as carelessness. Poor on-the-feet thinking and communication skills do not advance reputations. The mumbling absentminded scientist, a stereotype of fiction, has no place in modern corporate management.

Just as family members are told not to argue in front of strangers, so intracompany representatives must be careful to settle their differences appropriately. The place to solve a conflict between members of the same company is clearly not in an intercompany meeting. Differences can and should be acknowledged, if they are important to the overall effort, but the resolution is expected to take place "back home."

Intercompany meetings are extremely taxing in terms of time and effort, but they should be expected. Customers have the right to know what is happening in the projects for which they hold prime responsibility. At the same time, they expect all the technical personnel among all the different companies under contract to serve as problem-solvers. Common to intercompany meetings is the central discussion question of policy: How can each of our companies fulfill our contractual obligations for the improvement of the overall project?

International Variations

In 1984 *The Philadelphia Inquirer* published the global map reproduced as Figure 11-2. It represents only those large businesses located in the Philadelphia area and shows their holdings throughout the world. The map is extremely revealing, especially when you visualize a similar map for every other major city in the United States.

The entire world has come to be the business place of modern corporations. For the technical professional, interaction with peers

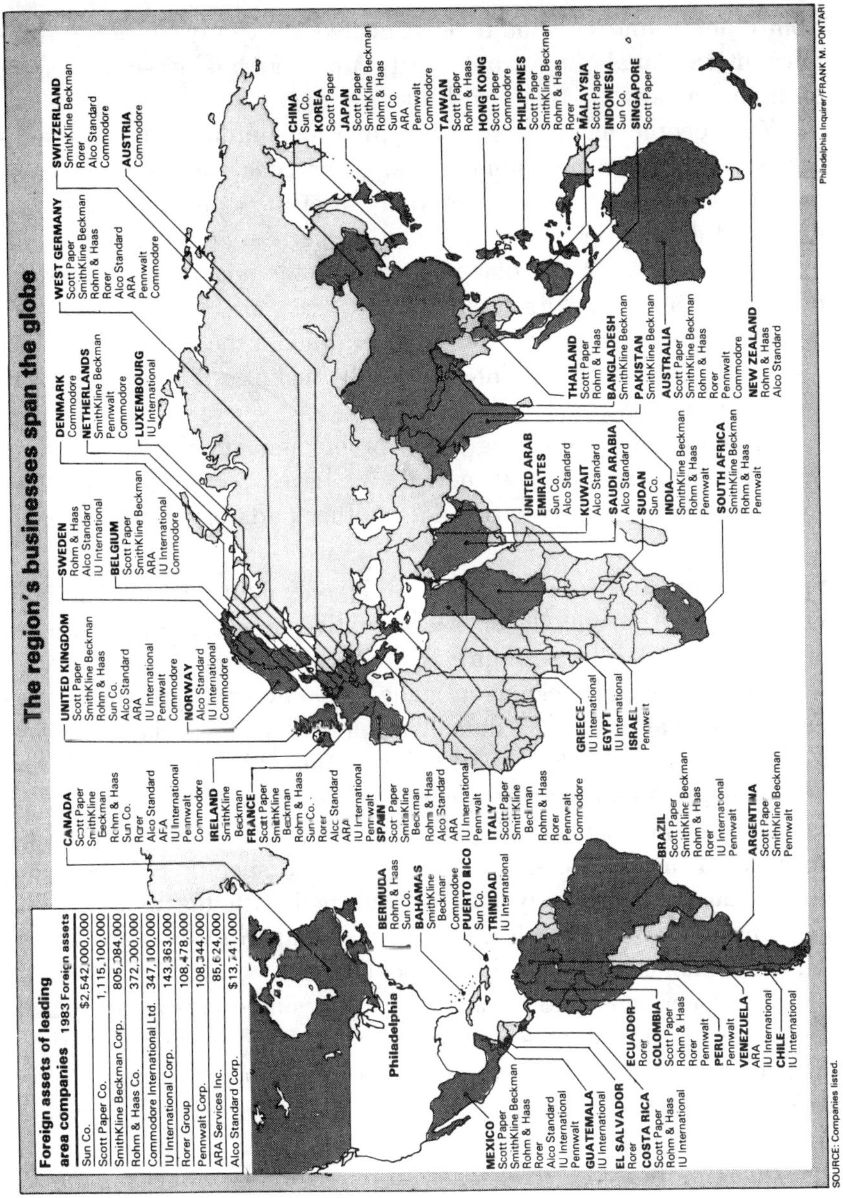

Foreign assets of leading area companies	1983 Foreign assets
Sun Co.	$2,542,000,000
Scott Paper Co.	1,115,100,000
SmithKline Beckman Corp.	805,084,000
Rohm & Haas Co.	372,000,000
Commodore International Ltd.	347,100,000
IU International Corp.	143,363,000
Rorer Group	108,478,000
Pennwalt Corp.	108,344,000
ARA Services Inc.	85,624,000
Alco Standard Corp.	$13,741,000

Figure 11-2 Map showing world business holdings by companies in the Philadelphia area. From *The Philadelphia Inquirer,* 30 April 1984. Used with permission of *The Philadelphia Inquirer.*

from other countries should be expected. Not only do many firms have international offices, but purely American businesses also work on intercompany contracts with firms abroad.

Most people have become sensitive to intercultural differences and how they harm communications. Companies sending representatives to other countries now routinely brief their people on the customs and differences to be expected in the host country.

This book is not the place to examine the spectrum of intercultural differences. We urge you to familiarize yourself with the expectations of any receiver you plan to meet. Should the time arrive when you are scheduled to work interculturally, make certain that your firm helps you prepare for the differences.

An extremely interesting group problem-solving trend operative internationally is worth your notice. We have observed it in parts of Europe. It is undoubtedly present in other parts of the world too.

Spokesperson Approach In 1971, one of us was working in France with Aerospatiale on a cooperative program. In the review meetings, similar in structure to those described in the previous section, a single spokesperson made the entire presentation for the French company. Rather than using a series of specialists whose staffs actually worked on a particular portion of the task, the French used one individual to present everybody's input. All questions to be asked were the responsibility of this person, as was providing all answers. If he did not know the answer to a specific question, the group caucused privately and he returned with the answer. The person who had the knowledge was not permitted to state it directly. All words came from this single spokesperson.

The Americans had to adjust to this single-person approach. Using the interactive model, they spoke when they wished. American speakers who did not know the answers to questions they were asked deferred to the specialist, who spoke directly. Typically, several Americans would comment on the point, and anyone who had a question simply asked it.

Our German sources claimed that they used this spokesperson approach infrequently, except when negotiating with governmental bodies. One did assert that he found it used routinely among the Japanese.[15] Perhaps the French and Japanese cultures lend themselves to the single unified-statement approach. We also suspect that, like the Germans, Americans use it when issues of sensitivity, such as labor-management negotiations, mandate the appearance of complete agreement.

We do not know how widespread corporate group variations might be, but our point is that a technical person must be ready to adjust to them. Developments such as the quality circle, for example, might or might not spread throughout American businesses. You must be ready for whatever changes come. Whether group process is easy for you or not, whether it is hybrid or according to expectations, highly formal or deceptively casual, modern business is likely to continue using groups for problem solving. The professional who is aware of group potentials will recognize that a career can be enhanced by those who participate competently.

WRAP-UP

Problem solving is the business of modern corporations. Groups provide the conducive environment that encourages solutions. From the amount of material we have presented here regarding this communication setting, you can see that we consider it important for technical competence. In business, the time spent in group settings is considerable. You must be able to analyze what is happening in a group and how you might contribute to effective solutions.

Not every collection of persons is a group, so our definition identifies required numbers, purpose, norms, cohesion, and consensus. Types of groups vary also, and we have focused on the problem-solving task group.

Group communication has disadvantages, but it has advantages so significant that modern businesses pay the costs. Advantages include the quality of solutions and their acceptance by participants.

Understanding the existence of different kinds of discussion questions is important for successful task groups. Questions of fact, value, and policy demand different sorts of group activity. Policy changes are the usual major concern of businesses.

When we had established the fundamentals of group structure, we looked at some modern industrial group models. Arbitrarily dividing the meetings into in-house and intercompany categories, we find that group structure is pervasive. The in-house progress review meeting is extremely important, and we discussed varieties ranging from the upper-management PAR business review to the specialized technical-unit meeting.

Finally, we looked at intercompany group meetings. These are usually quite formal, use an agenda, and are always carefully prepared. Two international variations of group structure warrant noting: the French and Japanese use of a single spokesperson and the widespread emergence of quality circles.

The technical professional faces a business environment where competence in group communication becomes vital. Whether in-house or intercompany, single-culture or international, the same conclusion is clear: a person who plans on career advancement must function effectively in groups.

ACTIVITIES

1. Identify any five or six associations of people with whom you gathered in the last month. Using the definition of group advanced in this chapter, were these associations of people a *group?*
2. Using any *group* you identified in Activity 1, list the disadvantages and advantages of meeting with the people. Which list did you decide outweighed the other? Would you be willing to meet with that group again?
3. For each of the topic areas below, write a *properly worded* discussion question of fact, value, and policy:

    - Increased research and development of military weaponry.
    - Addition of human interaction knowledge to standardized college entrance examinations such as the SATs.
    - Required teaching of basic astronomy in K-12 grades.
    - Government control over genetic engineering.
    - Use of human subjects in biological/chemical testing.

4. Using the corporate forms of meetings described in this chapter, examine which of these take place as a regular part of your campus administration/faculty structure. How do you account for any differences you find?
5. In the discipline for which you are preparing, what kinds of international cooperative meetings do you envision? What special problems would you forecast for such meetings? How would your future corporation have to prepare its members for such problems?

NOTES

1. Paul J. Rega, "The Art of Effective Persuasion," *IEEE Potentials* 3 (February 1984), 14–16.

2. Ernest G. Bormann and Nancy C. Bormann, *Effective Small Group Communication* (Minneapolis: Burgess, 1972), 6.

3. Halbert E. Gulley and Dale G. Leathers, *Communication and Group Process*, 3rd ed. (New York: Holt, Rinehart & Winston, 1977), 22–23.

4. Larry L. Barker, Kathy J. Wahlers, Donald J. Cegala, and Robert J. Kibler, *Groups In Process*, 2nd ed. (Englewood Cliffs, N.J.: Prentice-Hall, 1983), 8.

5. B. Aubrey Fisher, *Small Group Decision Making: Communication and the Group Process* (New York: McGraw-Hill, 1974), 31–32.

6. The authors found the use of an agenda for intercompany meetings reported as widespread in our research in Sweden and Germany. Interview with Ulf Hall, Bjorn Ankarsward, Olle Andersson, Arne Karlsson and Lars Wilhelmsson, Volvo Flygmotor, Trollhattan, Sweden, on May 16, 1984, and with H. Juntgen and J. Klein, Bergbau-Forschung, Essen, West Germany, on May 29, 1984.

7. George D. Demos and William F. O'Neill, "'Groupthink': The Study Group in Thought Reform," *Et Cetera* 41 (Summer 1984), 187–205.

8. Gulley and Leathers, *Communication*, 57–58.

9. For a discussion of a study that links decision participation with job satisfaction, see Lawrence R. Wheeless, Virginia E. Wheeless, and Richard D. Howard, "The Relationships of Communication with Supervisor and Decision-Participation to Employee Job Satisfaction," *Communication Quarterly* 32 (Summer 1984), 222–32.

10. Interviews with Gunter Alisch, Daimler-Benz, Sindelfingen, West Germany, on May 18, 1984, and H. Juntgen and J. Klein, Bergbau-Forschung, Essen, West Germany, on May 29, 1984.

11. The composition of the business review meeting is peculiar to a particular corporation, in this case General Electric. We suspect that some version of this "upper management" business review exists in most large companies. Those in attendance at these meetings would vary with the needs and traditions of the particular corporation.

12. William M. Ringle, "Made in Japan," *Nation's Business* (February 1981), 67–70.

13. Quality Circle Institute, *Quality Circles: Answers to 100 Frequently Asked Questions*, (Red Bluff, Calif.: Dewar Associates, 1979), 9.

14. Quality Circle Institute, *Quality Circles*, 6.

15. Interview with Hans Juntgen, Bergbau-Forschung, Essen, West Germany, on May 29, 1984.

CHAPTER TWELVE

SMALL-GROUP PARTICIPATION AND LEADERSHIP

The technical professional usually does not have the luxury of choosing whether to work alone or on a team. Groups are an inherent part of modern business. In Chapter 11, we dealt with the theoretical basis of group activity. In this chapter, we look at the actual group setting to help you develop an understanding of competent teamwork. We will discuss how to prepare for problem-solving groups, how to organize an effective agenda, how to run meetings according to parliamentary process, how to identify roles as they develop, and how to function as both an effective leader and participant.

FORMAL MEETINGS AND FORMAL PREPARATION

In the business world, some meetings—especially important ones—are highly structured. Participants, often from different companies, know what is to be discussed, in what order the issues will be introduced, and what action is expected to result. We call such meetings *formal.* They are inherently neither superior nor inferior to unplanned and unstructured meetings, which also occur frequently. Business operates with both sorts, but it is the formal meeting that consumes the professional's time in the sort of proposal contract reviews we discussed in Chapter 11. We will focus on such formal meetings here and assume that you will adapt whatever is useful to the more casual format.

Meeting Agenda

An *agenda* is a list of business items to be discussed in a group in the order they will be considered. In this sense, an agenda outlines the structure of a meeting by communicating to the members what will take place and in what order it will happen. Both within the United States and internationally, corporate meetings often function with an agenda; and, the more crucial the meeting is seen to be, the more likely it is that an agenda will be used. In the following paragraphs we list the advantages of using an agenda in the business setting.

An Agenda Gives Opportunity for Prior Thought Because group members know what the order of business will be, they can plan and form strategies before the meeting itself. Reports, data, and supporting documentation often accompany the distribution of an agenda. Participants, therefore, can prepare the background materials while being aware of the order in which these materials will be used.

An Agenda Gives Opportunity for Securing Prior Information When people think about issues that will be raised in a meeting, they can locate the areas where their own information is deficient. Having specific data in hand during a particular meeting can be the deciding factor in securing the support needed for a given position. Members of business groups are expected to have all pertinent information and documentation ready for their meetings.

An Agenda Forces Decisions about Meeting Goals to Be Made in Advance We assume that meetings are held for some specific purpose. Just what that purpose might be is not always clear to all members. Participants prepare for and expect quite different discussions and results when a meeting is convened only to explore the background of some problem, than when the purpose is to solve it. An agenda communicates the extent of the issues to be discussed.

An Agenda Helps All Members Keep Track of the Order of Business With the organization of issues before them in written form, even the most distracted members or digressing groups can find their place again. In many ways, the agenda works like a road map to tell you where you are at any particular time.

An Agenda Tells Members Where They Stopped and How Much More Is to Be Done Often a group will not accomplish as much as its planners had hoped for during the allocated period of time. An agenda helps participants to see how much more remains to be done. If the list of issues still to be settled seems longer than the time remaining, another meeting can be scheduled.

An Agenda Helps Members Assess the Productivity of the Meeting When people understand the goal of a given meeting, they can reflect on how successfully the goal was reached. Productivity is more than just "accomplishing the task." A group might rush through a six-item agenda and end up with hurt feelings and shoddy answers to problems. They would not be productive. Productivity refers to how well an *individual member* evaluates the content and feeling levels of any group.

Types of Agendas

Some meetings have informal agendas. These usually result from groups that meet often enough for members to know each other well. Such groups develop a system of "Here's the way we always work," and the structure—almost never written or articulated—becomes the agenda. The steps through which the group reaches its consensus are familiar to all the members.

Sometimes there is a "hidden agenda." Regardless of the stated reason for the meeting, participants sense that something else is really happening. Perhaps one or two members of the corporation decide to show the boss that their ideas are good reasons to grant them promotions rather than their opponents. The promotions will never be mentioned in the meeting, of course, but other members get the feeling that what is being said is about something else. Attendance at a meeting with a hidden agenda can be an extremely frustrating experience. In effect, the group is being used to solve a problem other than the one stated as its purpose.

Your agendas should aid your group overtly, for they can be extremely helpful. Without an agenda to point the way, group members have difficulty sensing where they are and how much more they have to do. An agenda is a formal way to communicate expectations for a group meeting to the members. When used as a tool that guides, it is a most effective aide to productivity.

Dewey-Based Problem-Solving Agenda

In the early part of this century, John Dewey published some speculations on how people think while problem solving. He called the process "reflective thought pattern" and theorized that, when people "reflect" upon some problem, they go through several steps in evolving the answer.[1] The method interested group-discussion theorists as a promising way to examine how groups work.

Some recent books on discussion theory no longer consider Dewey as the last word on how groups operate.[2] Nevertheless, Dewey's model is attractive because it is easily understood and put to practical use, even by those just beginning to examine group interaction. We have not reproduced his pattern exactly, but rather have based the following seven steps on his work.

When a group meets to solve a problem requiring action (which we classified in Chapter 11 as a question of policy), the members might prepare themselves by taking the seven steps we suggest. Further, when they actually meet, they might use the seven steps as an agenda to guide them. Such a task group is expected to reach consensus on *each* item before it moves on to the next. The danger in this process is that a group can lock itself into the agenda and, contrary to participant feelings, march doggedly from step to step. Intercompany meetings are a case in point: when several companies come together with a preplanned agenda and agreement regarding how they will proceed through it, there is a real danger of following the plan blindly. Ideally, Dewey or any other agenda form is developed to help the group, not to stifle its productivity. Used with skill, the Dewey steps are excellent beginning points for organizing content.

I. Definition of terms.
The reason for meeting must be explored first, and this reason is written into a properly worded discussion question. Each potentially ambiguous word in the question and any related terms must be defined exactly as the members plan to use them.
Why are we meeting and how is each crucial term of our discussion going to be defined?

II. Analysis of the problem.
The problem must be taken apart so that its various parts are understood before any suggestion for solution is made.
Why is this a problem?

A. Symptoms and effects of the problem.
What specific harms has this problem generated?

- B. History of the problem.
  What is the background of this problem?
- C. Causes of the problem.
  Where did the background generate a difficulty?
- D. Extent of the problem.
  How widespread is the harm?
- E. Criteria by which to judge solutions.
  What standards will be imposed on any solution selected?

III. Possible solutions to the problem.
Each possibly workable solution should be listed with all the details necessary for understanding it.
How might this problem be solved?

- A. Solution 1
  What are the details of Solution 1?
- B. Solution 2
  What are the details of Solution 2?
- C. Solution 3, and so forth.

IV. Advantages/disadvantages of each solution.

- A. Possible advantages.
  What benefits does this particular solution bring?
  1. Solution 1.
  2. Solution 2.
  3. Solution 3.
- B. Possible disadvantages.
  What costs does this solution entail?
  1. Solution 1.
  2. Solution 2.
  3. Solution 3.

V. Selection of the best solution.
After comparing advantages and disadvantages, one direction of action is probably shown to be superior.
Which solution solves the problem best?

VI. Implementation of Solution.
Since solutions do not magically materialize, group members must understand what steps will take place to set the plan into motion.
Who does what to get this solution into action?

A. Identification of the power structure.
Who needs to be influenced to get this action done?
B. Division of responsibilities.
Who is assigned what duty at what step?
C. Schedule of completion.
By what date does what have to be finished?

VII. Monitoring and fine-tuning the solution.
No solution can be left to function without double-checking it and making needed adjustments.
Once the solution is in place, what mechanism exists to watch and alter it, if needed?
A. The monitoring mechanism.
Who is responsible for watching the solution?
How often and under what conditions will it be monitored?
B. The adjusting mechanism.
At what point does the solution warrant alteration?
Who makes that decision?

With our additions and expansions, we find the Dewey pattern particularly suited to the technical business world. Notice the assumption that no solution is ever perfect. Each plan involves costs, some of which have nothing to do with finances. Reallocation of labor, retraining of personnel, and resistance to change are all possible disadvantages. Most changes involve money as well, and any disadvantages must be significantly outweighed by advantages. Why else would anyone want to try a new way of doing anything?

With this approach the solution is not considered automatic. Step VI forces the group to look at how the plan will be adopted. Good ideas sometimes die in committee because the power person cannot be convinced to put them into action. Groups in business settings must deal with such realities by reaching consensus on how the strategies work and who is responsible for them.

Finally, Step VII fits well with the assumption implicit in the scientific method that all solutions require adjustment. Just as scientists must be willing to alter the theories they have advanced, so scientists as group members must be willing to change a solution found in need of fine-tuning.

Taken together, these steps increase the probability that a problem will be thoroughly considered. As long as a group does not regard the pattern as a lockstep coercive mechanism, it can be a helpful tool. A discussion group armed with such an agenda can prepare itself with needed information and bolster its conclusions with required proof. It can decide how much it should accomplish in any given meeting period. It can check its own progress. Finally, it can use its own members to debug any shoddy thinking.

In many ways, a group discussion is like throwing ideas into an imaginary pot in the center of the group. Each idea gets stirred around and changed before each member accepts it. When the stirring is finished, the content of the solution may be unlike anything each member started with; but what remains in the pot is the property of the group. Each group member can claim a bit of it, yet the whole belongs to everybody.

Parliamentary Procedure

Most of you will belong to professional groups, and a few of you might advance to leadership positions in those groups. When that happens, we guarantee you a moment of utter panic. You will be handed a copy of your organization's bylaws and discover that you must operate *according to parliamentary procedure*. Not yet terrified, you will buy a copy of Henry Robert's *Rules of Order*, Alice Sturgis's *Standard Code of Parliamentary Procedure*, or any other of several modern parliamentary books. The panic will come after you discover that there are several hundred thousand ways to combine various actions legally and, as one wag said, a million ways to do them illegally.

Because we expect you to reach the top of your professions, we want to clarify some of the seeming mysticism about parliamentary procedure. At the moment you panic, reread the pages that follow.

The code of parliamentary practice organizes what can happen at a given meeting into a certain structure. It is concerned with who can do what at what particular time. It is the most common organizational pattern in existence for formal groups, usually set down as a required rule in the group's bylaws. As a result, both members and leaders of that group are tied to its rules. No action is legal when done in violation of the group's own rules. If those rules include the mandate for parliamentary procedure, meetings must follow its dictates.

Businesses often engage in sensitive meetings with other organizations. A company cannot permit a negotiation to deteriorate because of conflict over who should speak and how decisions are to be made. As a result, parliamentary procedure is often the agreed-upon way of circumventing such damaging conflicts.[3] Its use is far more frequent than most of us guess.

Fortunately, the procedures are not really mysterious. Most meetings require only a small portion of the actions detailed in Robert or Sturgis. In this section, we present some basics of parliamentary procedure, some of the most common uses for it in meetings you will attend, and a quick-reference summary of the most-used actions.

Bylaws Sometimes combined with a constitution, bylaws spell out the rights of and constraints on every member of an organization. Ordinarily, they list the purpose of the group, its leadership structure, the duties of the leaders, the rights of the members, the limits of power, and the way meetings will be run. Bylaws usually provide for their own amending process, but no action permits the bylaws themselves to be set aside. In short, no group can function in violation of its own set of formal rules.

Upon accepting membership, each member of an organization ought to become familiar with the group's bylaws or its agreed-upon rules. These rules serve as a safeguard against any action that might constrain an individual's rights. At the same time, they allow a member to study the latitude of choices available for legitimate action in a group.

The Agenda The agenda, discussed in Chapter 11 and earlier in this chapter, is very important to parliamentary procedure and its composition is usually mandated in the bylaws as a duty of the leadership. A common requirement is that agendas be sent to each member with sufficient advance notice of a meeting date.

The agenda, sometimes called the orders of the day, will list the business to be covered during a meeting period. The items must be brought forward for group action one at a time in the order printed. The chair cannot consider something out of order without the group's permission to do so. The chair and secretary share the responsibility for maintaining the order, but each member should insist on it.

Minutes The record of a previous meeting is called the minutes. They must be printed and distributed to members. The duty of the members is to see that minutes are accurate: all action should be recorded as it happened; no action ought to be added and none omitted. Errors that occur should be corrected. A formal motion usually signifies that the membership agrees to the accuracy of the minutes as of that date. ("I move that the minutes be accepted as read [or as corrected].)" These accepted minutes become the official record of the organization. They are passed on from leadership to new leadership, whose duty it is to review the predecessors' work.

Reports Reports are often mandated in the bylaws or ordered by the chair. Their order is listed in the agenda. Such reports might be simple statements summarizing the work a person or committee did. As such, they require no action. Often a report does contain a recommendation for something to be done. In that case, a formal motion to accept the report and its recommendation is made. The motion is handled as is any main motion.

Making a Motion Motion making sounds deceptively simple; however, there are four different types: main, subsidiary, incidental, and privileged. Their functions make the difference between them important. Each has a given rank, and the rank makes it possible to introduce one when another is being discussed. We will concentrate only on main motions.

Only one item at a time can be on the floor as legitimate business for action. A member can introduce another item only if the motion is of *higher precedence* than the one being discussed. The precedent motion bumps the existing one aside for the moment. When the precedent motion's business is settled, the group must return to the motion it was working on when the second one was introduced. The group always works on the item most immediately moved. This is the higher-precedence motion, and it must be acted on first. When settled, the group moves to the one it was working on just before the precedent motion was made.

Look at Table 12-1. It arranges the motions according to precedence. At any point, a motion of a higher precedence can be made, but one of a lower precedence is out of order. The table also summarizes other important features of each motion, such as its effect, its need for a second, whether it is debatable and amendable, and the sort of vote it takes for passage.

If you follow the table, you will see that a main motion can be made and, while it is being discussed, any of the higher motions or all of them in order can legitimately be made. The group then considers the last one moved (the one of highest precedent) and goes on in sequence until it works its way back to the least precedent motion. *No two motions of the same rank can be introduced, however, until the action on the first is completed.* At that point, there is nothing on the floor, and a member is free to move any motion at any level. Once

Table 12-1 Summary of Common Motions

Priority	Type of Motion	Effect of Motion	Debatable	Seconded	Amendable	Vote Required	Can Be Reconsidered
Increasing Priority	Adjourn	Closes a single meeting. *Sine die* closes final meeting.	no	yes	no	majority	no
	Recess: if no question is pending.	Follows main motion requirements.	yes	yes	yes	majority	no
	Recess: if pending	Intermission of action.	no	yes	yes	majority	no
	Lay on the table	Delays consideration until later.	no	yes	no	majority	no
Motions of Equal Rank	Point of information	Asks for information.	no	no	no	No vote. Chair provides requested action.	no
	Point of parliamentary inquiry	Asks for procedural information.					
	Personal privilege	States personal request.					
	Point of order	Gives a procedural clarification.					
	Close nominations	Stops the naming of candidates.	no	yes	yes	two-thirds	no
	Previous question	Stops debate and moves to vote.	no	yes	no	two-thirds	yes, with restrictions.
	Extend/limit debate	Provides for more or less debate time.	no	yes	yes	two-thirds	yes, with restrictions.
	Postpone to certain time	Delays to a specific point in meeting.	yes	yes	yes	majority	yes
Increasing Priority	Commit (refer)	Sends matter to committee.	yes, with restrictions.	yes	yes	majority	yes, with restrictions.
	Appeal	Disagrees with ruling of chair.	yes, with restrictions.	yes	no	majority in negative	yes
	Amend the amendment	Changes the amendment.	yes	yes	no	majority	yes
	Amend	Changes the motion wording.	yes	yes	yes	majority	yes
	Postpone indefinitely	Kills the motion.	yes	yes	no	majority	yes (affirmative vote only)
Motions of Equal Rank	Types of main motions: Accept recommendations of reports. Accept/correct minutes. Old Business. New Business.	Conducts the usual business for which group meets.	yes	yes	yes	majority	yes

there is a motion on the floor at whatever level, only one of higher precedence is in order.

The order of precedence is not difficult to learn. Many chairpersons and members use a diagram system or simple list to remind themselves of where the group is at a given moment. If a member erroneously makes a motion, the chair should declare it out of order. If the chair does not, another member should demand the chair do so at once.

Once you have learned the order of precedence, consider these other issues:

- Make each motion carefully, wording it precisely. It is best (and some bylaws require that members do so) to write out motions and hand them to the secretary when they have been moved. That way, the secretary has an accurate copy of the motion for the minutes.
- Motions are worded by saying, "I move that . . ." Wording such as "I make a motion that . . ." is incorrect.
- No motion can be amended more than twice. That is, a motion can be amended and an amendment to the amendment can be made. The action proceeds first on the amendment to the amendment, then on the amendment (if the amendment to the amendment was adopted, its change now becomes the amendment) and finally, the motion that was first moved (as amended, if the amendment was adopted.) In schematic, it would look like this:

  Original motion
  - Amendment to the original motion
    - Amendment to the amendment of the original
- There is no such thing as a "friendly amendment." Whether friendly or hostile, the group must consider the amendment, discuss it, and vote on it.
- Once the chair declares some motion as under discussion, usually by saying, "It has been moved and seconded that . . . ," it is no longer the property of the maker of the motion. Its maker, therefore, cannot withdraw it or alter it in any way. The motion belongs to the membership.

Safeguarding Rights Parliamentary procedure protects member rights and insures that nobody is permitted to do anything that violates the bylaws where those rights are spelled out. A few motions, all extremely high in ranking, make it their special function to protect a member. Two of these, outranked only by the motions to recess and adjourn, are the motions of personal privilege and point of information. These can interrupt almost any action and are not subject to a vote. They are statements by a member about personal rights. Personal privilege is a plea made to the chair about an individual difficulty such as not being able to hear the speaker or wanting to open a window to clear the cigarette smoke from the room. The chair is expected to act, if possible, on the request. Point of information is a plea to the chair for the member or the chair to provide some needed instruction such as what the motion being voted on actually means or whether the discussion is on the amendment or the amendment to the amendment. Again, the chair is expected to provide the information.

Blocking Motions The intent to block a motion can be done with any of the following precedent moves. When a member wants to prevent a vote at a particular moment, a motion can be made to:

- Postpone indefinitely. The action effectively kills the motion.
- Commit. The action sends the motion to some committee for further study. The committee is expected to report the action back to the group at some later time.
- Postpone definitely. The action is removed from immediate consideration but must be brought back at the time it was assigned definitely to be.
- Lay on the table. The action postpones the motion's consideration until later in the *same* meeting for groups that meet less often than quarterly and until the *next* meeting for groups that meet more often than quarterly. If the motion is not removed from the table within the required time limits, it dies.

Adopting Motions into Action Meetings run by parliamentary procedure approve action through the mechanism of motions. Basically, all motions must be voted upon by either written ballot, voice, or a standing count. The method is usually stated in bylaws.

Bringing the motion to a vote first requires ending the discussion about it. Parliamentary procedure refers to the discussion as debate and specifies which motions are debatable and which are not. Table 12–1 indicates the debatable status for the motions listed.

The right to debate a given motion is a serious member privilege. Parliamentary procedure assumes that all viewpoints must be safeguarded by insuring each member's right to speak. The motion is ordinarily not brought to a vote until the chair sees that no further comment is pending. The chair then says, "Are you ready for the question (vote)?" If members respond affirmatively, the vote is taken.

At least one action is available to the membership for cutting off debate. It is used when people are satisfied that enough has been said on the topic, and they are ready to vote. A member can say, "I move the previous question." If seconded, it places before the membership the question "Should debate on this topic stop?" Because it has the serious effect of stopping further comment, the motion takes two-thirds of the voting membership for passage. If the two-thirds agree, debate stops on the motion, and the chair goes immediately to the vote. Many members do not understand this action. Often, when people want to stop debate, they simply shout out, "Question! Question!" We have seen chairs take that as a demand for immediate vote and actually stop the debate. The action is illegal. The only way to deny a member the legitimate right to speak is to do so by the motion of previous question.

Whether a motion passes depends on the type of count it requires. Most require a majority (one-half plus one of the members present in the room and entitled to vote). A few are so serious, such as the previous question motion, that they require a two-thirds vote (two-thirds plus one of the members present in the room and entitled to vote). The secretary determines how many members are actually present so that a calculation can be made as to what vote constitutes passage.

Bylaws usually place one further restriction on the vote: quorum. A quorum is the minimum number of members needed to be present in order to conduct any kind of business (take any kind of vote). Sometimes, a careless group will consider a crucial item late in its meeting, long after most members have left. If a quorum is not present, no action can take place. The secretary should be certain of the member count for each motion vote. If the quorum vanishes, the secretary must notify the chair. In long meetings lasting all day, the actual numbers of voting members present in the room, even while maintaining quorum, might change with each vote. Thus, the number needed to pass a motion might be considerably less than it would be if the vote were taken when all members were present.

Whether you lead or simply attend a meeting, it is your duty to see to it that every group conducts itself legitimately according to its own rules. Parliamentary procedure is one organizational pattern of action that works to guide a group through those rules.

LEADERSHIP AND PARTICIPATION

We have been concentrating on the structure of the group. It is time to look at the human side of the picture. Discussion groups have two main constituents: (1) the ideas (subject matter) discussed as well as the structure of the ideas (agenda) and (2) the people themselves as they generate feelings. The socializing give-and-take of groups is an immensely complicated process. We consider three aspects of this socialization: roles, leadership, and member participation.

Roles in Groups

When behaviorists speak of a role, they do not mean a fake identity taken on to portray a false self. As Lawrence Rosenfeld points out, different aspects of everyday behavior require different roles. A person taking college courses might move from the role of consumer when purchasing gasoline on the way to class, to student when sitting in a speech communication course, to expert when presenting a speech on techniques of offshore drilling in that same class, to advisor when having lunch with a friend in need of advice, to parent when returning home to care for a child. Each is a role; each is real; each is part of the personal self. Rosenfeld defines roles as "sets of behaviors. We identify different roles by the various behaviors they require. Each role has its own set of behaviors."[4]

The concept of roles is an important one for understanding what happens with human feelings in a group. People in groups exhibit behaviors. As any one member acts (speaks, yawns), the others react. Since they all contribute to the creation of group, each member has an effect on the entire process.

Group theorists have assigned descriptive names to the roles people assume in group settings. We have divided these roles into positive and negative categories. The positive roles help groups reach consensus with good feelings, while the negative ones do not. No role belongs *absolutely* in either category. For example, the person who acts to soothe the feelings of another would usually be considered as behaving in a positive manner. If that person consistently diverts the group from its task by directing discussion to one or another member's feelings, however, the group becomes an encounter experience but loses sight of its original goal—a solution to a problem. Such diversion from an established goal would be considered negative.

We cannot prescribe a formula approach to defining positive roles. Examine the communications taking place, note the group's established goal, then watch the behavior of individuals. To the extent that any behavior helps the group reach its goal and maintains good feeling levels, that role is positive.

Positive Roles

- *Clarifier:* a person who says and does things that help the group sort out ambiguities. People understand a concept, a fact, or perhaps a proposal better than they did as a result of the clarifier's intervention.
- *Peacemaker:* a person who acts to mediate conflict. As a result of the mediation, those in conflict begin to resolve their perceived differences.
- *Tension reliever:* a person who communicates something that allows the group or some members in it to return to a lesser state of stress.
- *Gatekeeper:* a person who keeps track of where the group is in its process and keeps the group informed of where it should go next.
- *Information giver:* a person who has pertinent data, required evidence, needed knowledge, and is willing to supply it.
- *Information seeker:* a person who asks for needed data, evidence, and knowledge, thereby leading the group to consider such factors in the decision.
- *Record keeper:* a person who assumes responsibility for maintaining an accurate record of needed group actions.
- *Energizer:* a person who communicates something that acts as a spark of interest to the group.
- *Supporter:* a person who allies with another person or with a set of viewpoints to sort out where the strength of support is.

Negative Roles

- *Nonparticipant:* a person who keeps aloof from the interaction of the group, saying little or nothing.
- *Blocker:* a person who stops virtually every approach to consensus. Tends to say "no" often.
- *Aggressor:* a person who attacks another verbally, thereby increasing hostility and tension.

- *Monopolizer:* a person who uses more speaking time than is fair, when others also desire to speak.
- *Clown:* a person whose communication tends so much to humor that it distracts from the point the group wishes to discuss.
- *Unprepared one:* a person who has not prepared by thinking about the agenda topics and accumulating information. This person's communication tends to be superficial or contrary to fact.
- *Playperson:* a person who wishes to play with the process of the group. The goals of the group are ignored in preference to the individual's enjoyment.
- *Recognition seeker:* a person who uses the group to bolster personal ego. Communications are directed at gaining the group's admiration.

Each of these roles is familiar. Most of us have either enacted such behaviors or watched another who has. Multiple and overlapping roles are not uncommon; neither are momentary ones. A single act of verbal aggression may be upsetting for a brief time, but the action is smoothed over as the aggressor takes on the role of supporter or even peacemaker. The problem accelerates when a behavior is repeated and a pattern forms. Some groups realize, for example, that they have a monopolizer among them. Despite all defenses and strategies, the person continues to consume disproportionate amounts of speaking time. Strong measures are necessary to prevent the meeting from being turned into a lecture.

Leadership in Groups

Another set of behaviors that evolves in the group setting is called leadership. It occurs when one person functions as "the one in charge" either by assignment or process evolution. This concept deserves separate treatment from the wider discussion of roles.

We define *leadership* as the willingness to risk communicating in a way that has an impact and results in influencing the action of another. Please note that each of the positive and negative roles can qualify as leadership functions. Understanding what might be happening in any group often requires setting aside many popular assumptions regarding leadership. The verbal person, for example, might or might not be the one exerting influence. We have seen extremely verbal persons turn into group nonparticipants. Conversely, one who finds it difficult to speak out in the public setting might become the leader in a group.

Our definition recognizes that a leader takes on risk. It is, after all, simpler to remain silent than to speak and have the group react. The leader accepts the risk, communicates, and—as a result of that communication—makes an impact. This impact, in turn, influences another's behavior.

Development of leadership in a group, even when nobody has been preassigned the position, seems to follow a pattern. One member eventually decides that participants cannot sit there silently, says something, gets people moving, and a true group forms. Under these conditions, leadership often shifts from person to person as the need for influence develops. Sometimes leadership centers on one or more individuals, sometimes not. The point is that leadership is the result of patterns of communication that have a certain kind of effect. In the business setting, leadership is tied to position as well as capability. Meetings are called by a supervisor, and the supervisor functions as the initial leader. By virtue of personal expertise over subject matter, others may take over leadership. At the meeting's end, however, business courtesy demands that the role be once again relinquished to the superior who closes the session.

We will first divide leadership into categories of assigned and emerged. Then we will consider how styles of leadership affect the process.

Assigned Leadership Sometimes, to paraphrase the quotation, greatness is thrust upon a person. There you are, chair of a campus committee or head of a corporate project. By virtue of your position, you know that you will have to lead the many meetings your task entails. Long before, immediately prior to, during, and after the group interactions, certain responsibilities are yours. We note twelve of them.

Plan the agenda. You must decide how much can be settled, which issues will be discussed, and the order of those issues. You will also need to decide the form this discussion will take. Will you use parliamentary procedure, some amended form, or no formal process structure at all? How will decisions be made: Vote? Private reaction? No recorded feedback at all? How will you handle conflict: Encourage it? Smoothe it over? Squash it?

Plan resources needed for the meeting. If visual aids are needed, as they would almost certainly be during corporate meetings, make certain they are prepared and machinery is in place to use them. People who might act as resources need to be contacted and given instructions. If these people are expected to attend the meeting, invitations must go out quickly.

Contact participants. Make certain that everyone who needs to be present is reached to check meeting times and availability. Telephone or in-person contact is a good idea at this point.

Prepare and distribute materials. Make certain that everything needed for background information is ready. Along with any materials you want members to read before the meeting, send the agenda. Members then can see how much is expected to happen and how much they need to prepare. Members also need to know how they will proceed. If parliamentary procedure is to be used, they need notice for preparation. If members are expected to present a brief (or long) statement, each must know this too. Written reminders of the time and location of the meeting, as well as a contact phone or office number, should be sent.

Provide a statement of expectations and goals. A good way to start the actual interaction is to clarify what is expected to happen. The first step of the Dewey process allows the scope of the meeting to be settled immediately. The leader should take the responsibility of stating initially what will happen.

Insure fairness of opportunity to interact. The leader often must act as traffic cop to hold back the people who use up too much time and to wave along those who are hesitant. Assume that each member is there because of a potential personal contribution to the solution. Make certain that nobody's contribution is overlooked.

Help the personal relationships. Not everyone in a group necessarily likes everyone else. Individual feelings have to be safeguarded in order to reach the established goal. Ignoring those feelings can jeopardize the group's progress. Leaders must be sensitive to what is happening at the human level. Bad feelings can develop into troublesome interactions and retarded productivity.

Aid the move toward consensus. Consensus, after all, is the aim of most problem-solving discussion groups. Leaders must do everything they can to see that their groups reach it. Some of their duty is procedural, such as appointing a recorder to keep accurate notes, securing needed information, or finding a structure that works. Problems with the group might be personal, and then the leader must handle whatever human feelings surface. Whatever gets in the way of consensus is acting as entropy. As with all communication losses, such entropy must be controlled.

Provide internal summaries. Groups sometimes move very quickly. It is easy to lose track of what has been done, especially if there is no recorder. A leader should make certain that each person understands where the group is on its road to consensus. Brief internal summaries such as "So we have decided that Frank's approach looks good, but we need information on the budget for that second stage, right?" are extremely valuable.

Provide closure. Closure is a concluding process. Leaders not only must communicate that the current meeting is ended, but they also must provide an overall view of what happened. This view serves as a final summary and provides the meeting with a satisfactory ending. Most of us have left a frustrating meeting wondering, "What was that all about?" Leaders should not condone such a feeling.

Clarify what must be done next. Beyond closing the current meeting, members are entitled to know what will be the next step. Leaders should not allow people to disperse until they know:

- Whether another meeting needs to be held.
- When and where that meeting will be.
- Who has responsibility for any assigned tasks.
- What deadline is expected for any assigned task.
- What information still needs to be accumulated.
- Where that information is and who is to gather it.
- What happens next if no meeting is scheduled.

Follow up on assignments and promises. People are easily caught up in many assignments. They forget statements made during a given meeting. Promises to do something are also forgotten. Leaders make certain that reminders prevent such losses. Leaders do not allow good ideas to become lost because there is no follow-through.

Emerged Leadership At times, the person who leads is not the designated leader. A group member arrives at the meeting place prepared to participate. Once the process begins, this person sees problems emerging and communicates in order to forestall those problems. The communication has impact and changes the behavior of another, thereby changing the behavior of the group. This person has acted in a leadership capacity. If this impact and influence continue into a pattern, the person emerges as the group's leader.

The above scenario is extremely common. The better the group, the more likely it is that emergent leadership will develop. We assert that, unless a group never actually gets itself going, leaders will surface. Assigned leaders are different, however, from emerged leaders. An outsider might choose the first, but only the group chooses the second. Whether assigned or emerged, leaders assume many of the same responsibilities.

Looking at the preceding list that describes the activities of the assigned leader, you begin to realize that emerged leadership can begin at almost any point. It may be as early as the first few words spoken within the group. Suppose that the group is brought together by some mechanism but is unsure or unclear as to what it will do. That very ambiguity might trouble one member enough to risk saying something about it, thereby starting the group thinking about its goals and constraints. As things progress, this same person might start providing information, clarifying, dealing with ruffled feelings, working to diminish tension, and keeping the group on course to its consensus. At some point, almost never openly expressed, the group begins to turn to this person whenever it feels unsure. Members make eye contact most often with this new leader, whose viewpoints hold more sway than do those of anyone else. To its own amusement, a group might find itself actually asking permission of the new leader to speak. As one of us remembers, a young man raised his hand to secure the attention of an emerged group leader. "I don't know why you have the right to tell me it's okay to speak," he mumbled, "but somehow if you say I can't, I won't." The person so addressed was stunned. In the post-meeting critique, she said that she had never consciously worked for the leadership. Things just needed doing and nobody else did them. She took the responsibility and ended with the given authority to regulate speaking rights.

Often, emerged leadership does not center on one individual but is shared by each member of the group. In an unstated democratic manner, those who can best do something for the group are given the job of doing it. Thus, the responsibility of impact shifts from person to person. Whether a group selects a single leader, a few leaders, or equally divided leadership, the function of the leaders remains the same: they alter the direction of their group's behavior.

Many of the tasks that the assigned leader performs are also performed by the emerged leader, some extending beyond the actual meeting time. An interesting phenomenon is the transformation of the emerged leader into the designated leader for a future meeting. We urge caution in grasping leadership in the business setting. Official position, as we have said, might make certain tasks a prerogative of that superior. Assuming those tasks against corporate traditions would probably be seen as presumptuous.

Styles of Leadership Remember that, whether assigned or emerged, leadership roles consist of behaviors. Different people have different ways of expressing their leadership. Theorists have called these differences *styles*. In an extremely influential study, White and Lippitt identified three types of leadership styles.[5] These have been widely discussed in group-leadership courses, and we adapt them below.

Heavy-handed leadership (authoritarian). Some leaders act with such directedness that nothing happens unless the leader permits it. Such persons stop members who seem to deviate from the "approved" course, regulate behaviors tightly into some predetermined pattern, tolerate no rebellions, and generally keep everyone in line according to personal rules.

Groups led by such people tend to:

- Lose the ability to function without the leader.
- Come to predetermined answers.
- Work to please the leader.
- Function with efficiency.
- Evolve a great many answers.
- Vary in the quality of their solutions.
- Develop high levels of frustrations.

Hands-off leadership (laissez-faire). Some persons do lead, but the leadership is so invisible that the members feel themselves to be leaderless. Such groups work out problems without guidance, flounder with difficulties, grapple with content and direction. The leaders permit this inefficiency or purposely place the members in these conditions. Maximum freedom exists under the hands-off leadership style. Groups such as this tend to:

- Feel unproductive.

- Evolve ambiguous solutions.
- Develop a free-flowing process of expression.
- Evolve their own rules and regulations.
- Lack efficiency.
- Develop great ego vulnerability for each member.
- Concentrate on feeling levels.
- Develop varying quality of solutions.

Evenhanded leadership (democratic). Many theorists see this style as the most beneficial. It falls between the other two extremes, by providing direction but never mandating it. Evenhanded leaders do not allow groups to flounder, but neither are they overly directional. Groups led by someone with an evenhanded style tend to:

- Become sensitive to member feelings.
- Develop multiple directions for solution, but confine them to consensus.
- Have fairly equal access to communication time.
- Like their group and its leader.
- Have varying efficiency.
- Balance productivity between subject matter and member feelings.
- Develop flexible rules.

We do not believe that any style of leadership is inherently better or worse than any of the others. There is obviously a time for the heavy-handed type, just as there is a time for the hands-off type. When the supervisor demands a quick answer, and time is short, a hands-off group will not meet the deadline. The heavy-handed group might. On the other hand, if you convene people from the assembly line to discover their feelings and frustrations, the heavy-handed leader will learn little. The hands-off leader who listens probably will. A good rule of thumb is that the higher the demand for content answers and the lower the concern for individual feelings, the more likely it is that the heavy-handed approach will work. The less the concern for content and the more the concern for people, the more a hands-off style works. Basically, the three styles form a continuum:

High content/Low feeling concern—Balanced concern—Low content/High feeling concern

Heavy-handed	Evenhanded	Hands-off

Seldom does a crisis situation demand a solution at the price of all participants' feelings. At the same time, seldom do solutions matter so little that only member feelings are important. Good discussion groups recognize the realities of both components. A leader who keeps the group working on answers, but allows freedom enough for beneficial feelings to emerge, does not get answers with maximum efficiency, but does retain the good will of those who participate.

The question of which style is best involves a trade-off between the immediate and long-term goals of the communication. Just as a speaker does in preparing for an audience (see Chapter 6), a group leader must keep in mind the eventual goal as well as the immediate one. Group leaders motivate to action. If their goal is an immediate answer at the price of an eventual deterioration of the working staff, they must have full knowledge of what they choose. If they spend all their resources developing the immediate feelings of their members and the price paid is incomplete answers to work problems, they must also realize the implications of this choice.

Leadership and personality is yet another way of explaining style. We divide leadership into heavy-handed, hands-off, and evenhanded by watching what a leader does during a meeting and how the group responds. This theory does not speculate on why a particular leader follows a certain style. Some have argued that the *why* of action is the most important consideration. Since the early 1960s, many organizations have used the Myers-Briggs Indicator to explain leadership styles.[6] Myers and Briggs, a mother-daughter team of psychologists, claim that behavior is tied to personality. Individuals act in certain ways because of the sort of persons they happen to be. The specific categories of individual personalities are measured by a machine-scored multiple-choice test. One type of person might be "down-to-earth," while another is "heads-in-the-clouds." In groups, one person might usually behave as "decided and fixed," while a fellow group member might be at ease only with a "be flexible and gather-more-data" approach.

It might be extremely interesting for you to take the Myers-Briggs Indicator or one of the shorter amended versions.[7] Understanding yourself is important, and such tests might help you explain your group behaviors. No behavior, however, is absolutely unchangeable.

Training and education permit people to alter their actions, including those attributed to personality. No matter what kind of leadership a Myers-Briggs measures for you now, you have every reason to believe that it can be changed, honed, and developed as you would wish.

PARTICIPATION AND RESPONSIBILITIES

If each group member took on responsibilities for group interaction in the shared system of emerged leadership we discussed earlier, there would be an ideal world of multiple group leaders. Every member would perform the tasks we listed as the jobs of leadership. Meetings would be productive in terms of both content and positive feelings. People would take on the needed positive roles and discard them when they were no longer of use. Such a utopia is not likely to exist, so each of us must prepare for less-than-perfect member participation. In this section, we call your attention to the concepts of cooperation, negotiation, preparation, and responsibility.

Cooperation and Consensus

A discussion is not a debate, although debates might take place in them. Neither is a discussion a series of arguments to force acceptance of preconceived solutions, although groups experience these pressures regularly. Discussion is about cooperating with others in order to stir up a community pot of ideas.

The assumption underlying any meeting is that multiple viewpoints are useful. In Chapter 11, we postulated that ideas generated by a group may yield better solutions than those arrived at alone. Whether that assumption is accurate or not, the modern approach to problem solving in daily life as well as in business is through use of a group. Stated another way: however you feel about them, groups are a necessary part of your life.

If we assume that multiple viewpoints are superior, we should state an important corollary: the ideas of the individual must be subjected to analysis and change. All the people walking into a Tuesday morning meeting may feel that they have the best answer to the pending question. The more these people are locked into defense of their own answers, the more likely it is that the discussion will fail. Group members using one another to find answers must be willing to listen, evaluate, alter, and accept other ideas.

Cooperation does not mean blind acceptance and an absence of challenge. Any idea surfacing in any group should be analyzed from all points of view. Good ideas are hard to eliminate when group cooperation is working well because they prove themselves superior when group members submit them to scrutiny. A story is told of Thomas J. Watson, chairman of the board of International Business Machines. In 1943 Watson did not agree with his staff about the promise of a novel machine being developed—a machine called the computer. "I think," he told them, "there is a world market for about five computers."[8] Fortunately for IBM, the staff position prevailed.

The same group scrutiny is true of poor ideas. When weaknesses surface, the people sitting together might see them more easily than a single person would. Someone else's experience could help change a viewpoint or even correct the flaw. Working together in the group spirit of cooperation, each member must be willing to yield and accept portions of an answer. This spirit also presumes that the final answer is the group's product, not that of any single member.

Negotiation

Cooperation and its product, consensus, involve a great deal of give-and-take. In an ideal situation, group members do not browbeat each other, do not resort to coercion, and do not maintain inflexible positions. What they do is negotiate an answer so satisfactory to themselves and the others that each can live with it.

Negotiation is the verbal process of giving up something in order to demand something else. It always involves give-and-take, a willingness to meet halfway.

In no way does negotiation mean *giving up*. Surrender is not a part of negotiation. Each member in the process must feel that what was obtained is worthy of what was relinquished. If they cannot, then consensus for them is not possible. They are at stalemate.

In his influential book on interpersonal communication, John Keltner says of negotiation:

> *Both parties must be willing to discuss the situation with each other in "good faith." While the definition of "good faith" is always elusive, the term is generally accepted to mean that the parties are willing to and intend to sit down with each other and that they expect to reach a mutually acceptable solution.*[9]

Although Keltner is speaking of formal negotiations, we contend that this same spirit of "good faith" must be present in any successful group. If participants enter with the attitude of cooperation and the expectation of give-and-take, chances of consensus improve.

Preparation

By now you probably know what we will say about these participant duties. It is what we have said about all communication:

- There is no shortcut.
- The responsibility for a communication belongs to all its participants.

Because there are several participants in a group, some people feel they can hide their ignorance. Untrue! Lack of preparation makes you vulnerable to those who have done their homework. The contrast between your level of knowledge and theirs becomes obvious quickly. You lose credibility, and your contribution to the answer invites problems. In addition, your lack of specific knowledge makes you unable to analyze someone else's idea. Most ideas and solutions seem acceptable or workable at first glance. When you have learned a bit about them, their flaws surface. If you are not prepared, you cannot know where to look for those flaws. Prepare! If you do not, you are a danger to your group and a disappointment to yourself.

Your responsibility, in short, is to do all that you can to help the group communicate well. Where leadership is needed, provide it. Where negotiation is required, do it. Work with others for the best answer to the group's problem. In exchange for their work, you offer yourself. Together, you can arrive at that group-generated consensus.

WRAP-UP

Participation and leadership constitute group activities. The theory we discussed in Chapter 11 helps you understand the interactions discussed here. We have concentrated on leadership and participation in formal group structures because most important meetings follow these patterns.

Agenda formulation has special advantages in formal meetings and is a widely used procedure in industrial meetings. We traced preparation for problem solving through the Dewey method and explained parliamentary procedure as a widespread method of group function.

Groups deal not only with the content of problems but also involve personal feelings. Participation in groups involves more than the discussion of subject matter. Participants behave in certain ways, and theorists give such behaviors the name "roles." All group members have roles, so all participants must deal with them.

Leadership and group participation are different aspects of the human-relation side of groups. Leadership might be assigned or it might emerge. A leader's style might be heavy-handed, hands-off, or evenhanded, but some sort of leadership always exists when you have a group as we have defined it. Leadership style might be due to the sort of personality members have, so the Myers-Briggs approach was suggested as a way of classifying people types.

We see participation overlapping with leadership in an effective group. Ideally, all members should strive to take responsibility for helping the group reach its goal. Suggestions for effective leadership and member duties remind us that the process of group discussion is never simple. Practical experience in the technical business world tells us that it is extremely important.

ACTIVITIES

1. Prepare a written Dewey-based outline for any of the policy discussion questions you wrote for Activity 3, Chapter 11.
2. The class should divide itself into groups of no more than seven students. Each group selects a topic area and writes a properly worded proposition of policy for that topic. Each member of the group prepares a written Dewey outline for his or her own use. The group comes together and discusses the topic (with the class observing). The Dewey outline agenda must be followed. At the conclusion of the allotted time period, the class observers should evaluate:

    - Did the group follow the outline pattern?
    - At what points (if any) did they deviate from it?
    - Did the pattern seem to help or hinder the group's progress?

3. Divide the class into two fairly even groups, called Group A and Group B. Select a topic of some controversy. Group A will be *in support* of some action concerning the controversy. Group B will be *against* the action sought by Group A. In a meeting run by

strict parliamentary rules, Group A will attempt to get its action into a main motion and approved by the body. Group B will do everything legal to stop the motion from passing. Appoint a parliamentarian and a chairperson.

4. Design an evaluation form that measures the skill of a group discussion leader. During a class discussion (or at a campus meeting outside of class), use the evaluation form to assess the performance of the leader.
5. Using the same procedure as in Activity 4, design and use an evaluation form for group-member participation.

NOTES

1. John Dewey, *How We Think* (Boston: D. C. Heath, 1933), and John Dewey, *Logic: The Theory of Inquiry* (New York: Holt, Rinehart & Winston, 1938).
2. Halbert E. Gulley and Dale G. Leathers, *Communication and Group Process*, 3rd ed. (New York: Holt, Rinehart & Winston, 1977) and Rudolph F. Verderber, *Working Together* (Belmont, Calif.: Wadsworth, 1982), for example, do not mention the Dewey process at all.
3. During meetings of the Solar Industries Association from 1977 to 1979, one of your authors participated in the formulation of the Solar Energy Collector Rating Standards. Many companies were involved, and the group met throughout the country during those years. At the initial meeting, to forestall quibbling about procedure, parliamentary procedure was agreed upon as the means of arriving at decisions. It was used throughout the duration of the task.
4. Lawrence B. Rosenfeld, *Human Interaction in the Small Group Setting* (Columbus, Ohio: Charles E. Merrill, 1973), 102.
5. Ralph White and Ronald Lippitt, "Leader Behavior and Member Reaction in Three 'Social Climates,'" in Dorwin Cartwright and Alvin Zander, eds., *Group Dynamics*, 3rd ed. (New York: Harper & Row, 1968), 319.
6. Isabel Myers, *Manual: The Myers-Briggs Type Indicator* (Palo Alto, Calif.: Consulting Psychologists Press, 1962).
7. David Keirsey and Marilyn Bates, *Please Understand Me: Character and Temperament Types* (Del Mar, Calif.: Prometheus Nemesis, 1978).
8. "Quote of the Century," *Academic Computing Services News*, West Chester University, (13 December 1984), 4.
9. John Keltner, *Interpersonal Speech-Communication* (Belmont, Calif.: Wadsworth, 1970), 240.

CHAPTER THIRTEEN

PUBLIC COMMUNICATION

For many people, speech is synonymous with *public* speaking. The intrapersonal communications, nonverbals, dyads, small-group meetings, and other ways people send messages orally to each other do not come to mind so easily as does the usual picture of a *speaker*—one individual up on a platform addressing a hushed audience. That picture is sometimes accurate, for humans often do use this type of public speech, but we hope you recognize that this type of communication setting is only one among many others.

Technical professionals are quite often called upon to be public speakers. In previous chapters we have shown how businesses use professionals as prime sources of messages to sell ideas or products. It is now time to deal directly with the settings that place these people on the public platform. We will see how many of the suggestions we have already made might be put to use in the public speech. Our thesis is that ambitious and successful technical professionals inescapably end up in public settings. Such situations include meetings of professional organizations, various sales and review presentations, public relations community-speaking assignments, and the modern demands of television. In each of these, competence in the special demands of public speaking is both expected and rewarded.

We define the *public communication* as any communication in which one person is primarily responsible for acting as the source of oral messages for an extended period. The listeners, more than two people, suspend their interaction with each other in order to remain fairly silent until the speaker has concluded. Using this definition, the public speech shows up in a variety of settings: a scientist reviewing test results in a corporate meeting, a teacher lecturing a class, a sorority president explaining the roles of her organization to the new pledges, or an uncle telling his relatives the long story of past family exploits.

Neither a stage nor any number of listeners beyond two are part of the definition of the public speech. The important factors are (1) a communication beyond the dyad, (2) responsibility for the encoding being centered in one person, (3) a message expected to last a protracted length of time, and (4) listeners who expect the speaker to carry the message-sending burden.

A communicator can learn much about successful presentations by reviewing the long history of public address. The "art of public speaking" has been studied and argued for nearly three thousand years. When Aristotle wrote what many consider the first full treatment of speech in *Rhetoric* around 300 B.C., the topic already had a long history. Aristotle, like Cicero and Quintilian after him, focused on the type of formal communication that came to be called oratory. By the nineteenth century, oratory and the eloquent speaker were much admired for their entertainment value. This appreciation of the form also solidified the notion that public speech meant a formal platform setting. Long and flowery speeches came to be regarded as an art form for their own sake.

Gary T. Hunt states that the twentieth century saw a shift of focus from the "speech as entertainment" to the "study of communication." "Instead," Hunt notes, "it [public speaking] is studied as a useful tool for communication among people."[1]

It hardly seems necessary to note that the effective use of public speaking can bring results. Speakers as different as Jesus of Nazareth and the latest American presidential candidate have understood the need to speak ideas so that an audience understands them. The technical professional recognizes this need, for the business world uses public communication far more frequently than most people expect. Wilson and Arnold, giving the term "public speaking" much the same meaning we do, state

> *We are going to focus on a kind of speaking that is commonplace for educated, professional people. They speak* in public *when they report, give instructions, talk to committees, support ideas in small or large groups, or talk to entertain others.*[2]

Because technical professionals are often required to use the public-speaking option, they must understand this special setting. Whether in-house or intercompany, at a professional organization or a monthly meeting of the township's library study group, educated people are expected to present their technical expertise competently.

As Wilson and Arnold note, "there is a great difference between 'getting by' in talking relationships and understanding how to make the most of them."[3]

FOCUSED VARIABLES

The speaker engaging in public communication uses all the variables we discussed in Parts I and II of this book. The differentiating factor in the public setting, as opposed to dyads or groups, is that one person, the speaker, has prime responsibility for sending the message forth. The audience, on the other hand, is hardly passive, for the message is being re-created and reacted to in each individual's mind. The assumption of public communication, however, is that the audience is willing to give up the prime sender's role and will remain fairly silent until the speaker has concluded. To understand how some of the variables work under these conditions, reexamine the following concepts.

Formalized Goal

We spoke at length about communication goals in Chapter 6. There we postulated that a speaker communicates to satisfy personal ego, to motivate another to change a viewpoint, and to effect action. These three modern goals incorporate the classic categories of speaking to persuade, convince, inform, entertain, and inspire. We think that each of these classic goals involves an intention to shift viewpoints or behaviors. Each also allows the speaker to achieve significant personal ego satisfaction.

The public speech begins with a goal. That goal must be achieved within a finite time, usually only the brief minutes of the speech itself. The speaker enters the situation knowing that a modern audience's attention span is very short and so accepts the duty of sending a message designed to achieve the recognized goal within that brief time.

The effect of a given speech certainly might change radically over the years. Many great speeches such as the Gettysburg Address were not immediately received with the acclaim they later achieved. In 1968 Fredrick Trautmann warned that it was impossible to establish cause-effect relationships between a speech and any set of effects attributed to it.[4] He is probably correct. Speakers leave platforms thinking they have failed to reach their goals and find years later that their speeches are considered successful.

In this sense, speakers have a conscious recognition of an opportunity to accomplish something in the span of speaking time. Whether they achieve it is quite another question.

The Specific Audience

Recall from Chapter 6 that a good communicator tailors messages to particular listeners. This "fit" applies to public settings as well. Because public messages often take place in front of many persons, speakers must be aware of the differences among individuals. A single speech is unlikely to satisfy everyone in a large audience and rarely will do so even in a small one. The effective speaker must accept the loss of some listeners. Learning to adapt material to reach the largest possible portion of the audience (or to reach a carefully selected portion) forces the speaker to think in terms of others.

The Nonverbal Messages

Any group of people develops certain conventions and traditions. Many of these are nonverbal. Speakers are expected to dress and sound a certain way, to maintain eye contact with some directness, to move and utilize space effectively, to use facial expressions appropriately, to present visual aids with skill, and to control the body with self-discipline.

Audiences do not expect these nonverbals to be identical, and not all speakers perform them in the same way. We have insisted that a formula approach to each speaking situation is impossible. Statements that claim "All speakers must gesture in the following way . . ." or "All audiences want . . ." ought to put you on guard. A careful audience analysis should guide you to select those nonverbals appropriate for you before a given audience.

Technical professionals appear before many audiences. When representing a company in a business climate, they are expected to satisfy the listeners' expectations of business dress, composed demeanor, and use of visual aids. The same people attending a conference in a resort setting can alter dress to some informality, but the "business posture" must still be maintained. In the final analysis, nonverbal messages tell the audience who you think you are and what you feel about them.

How the audience's expectations can be mistaken is demonstrated in the true example of a politician arriving on campus to seek student votes just before a local election. He was a portly middle-aged man who, erroneously trying to fit in with his youthful audience, wore tight jeans and a university T-shirt. Even if one were willing to forgive his substantial paunch, his attempt at using a slurred, casual articulation made him appear ridiculous. Sensing insincerity, the students reacted badly. They expected an overweight middle-aged man who sought to represent them to present himself differently. The candidate's verbal message was virtually ignored, because the entropy of his nonverbals was so strong. Such mismatches of audience expectation and speaker reality can often be traced to the speaker's choice of inappropriate nonverbals for the audience.

The Verbal Messages

Chapter 3 centered on the pitfalls of language, both verbal and nonverbal. The technical person must understand the impact of words but confronts an additional difficulty. Scientific and mathematical disciplines develop highly specialized vocabularies. These words make sense to people trained in the subject matter, but their use can be disastrous in a public presentation to lay persons.

Every discipline develops its own jargon. The specialized words sometimes enter the vocabulary of the general public. The development of computers has provided an interesting language case study. Words such as "software," "program," and "floppy disk" are used by those who have little knowledge of computers, but specialized jargon such as "REMs," "Bits," and "EPROMS" remain unfamiliar. Anyone using unfamiliar words with a lay audience has a special problem and a communicative duty.

Difficulty with jargon goes beyond danger that the message will be misunderstood, although that is certainly a serious source of entropy. Listeners often suspect that the jargon-user is hiding behind the words or talking down to them. Neither perception helps a speaker's credibility. The stereotype that "You can never understand what scientists are talking about" probably stems from those speakers who have remained oblivious to the effects of their language use. Treat technical terms with care. If you cannot find synonyms, supply definitions. If you discover that you *are* hiding behind the jargon in order to enhance your feeling of superiority, you need to have a harsh intrapersonal talk with yourself.

Make certain that the level of the language is suitable to the audience. Unlike our paunchy politician, a speaker does not have to talk "just like" the audience but does have to make certain that what is said can be understood. Again, the vital audience analysis should tell you what kind of listeners you have before you.

Choices of Delivery

The public speaker can choose from at least four ways to deliver the speech. Conditions, personal habits, and audience expectations will help determine which to choose.

The manuscript speech is one in which the entire speech is written out ahead of time and read to the audience. It is probably the most difficult speech to do well, for very few of us are such skilled actors that we can read a page and make it come alive with sincerity. The only time to make such a dangerous choice is when prior copies of the speech are required for the record. Presidential speeches are often handed out to the press hours before the actual delivery. Such action is perhaps justified by the need to get the speech into the newspapers and to have it become part of a historic record. Most of us watching the last few presidents attempt to read their manuscripts with some semblance of skill know that the read speech seldom succeeds. If presidents, most of whom have broad communication training and are seasoned platform speakers, cannot carry off the manuscript speech well, inexperienced students and technical professionals should be wary.

A significant problem with the manuscript speech is that it ties a person to a given flow of words. Most people deviate little from a prepared message. Feedback is lost and, regardless of how the audience responds, the speaker tends to plow ahead with the prepared words. Losing one's place and enduring an embarrassing shuffle of papers is still another problem.

The manuscript speech is used widely in a professional society's meetings. One "presents a paper" at the convention, and to many people that means all papers will be read. Some conferences even require that the participants bring copies of their paper with them for distribution to attendees. With such encouragement, the manuscript speech appears to be the easiest to prepare, for the paper itself already exists.

After watching many such presentations in both the United States and Europe, we wonder if the security the manuscript speech provides is worthwhile, for papers are often read with great ineffectiveness. Audiences sense the speaker's lack of contact and they react by showing no interest. We think the audience would be better off to simply take the copy and read it at leisure than to have the author read it aloud, barely seen, from behind a huge podium. This "talking head" with the droning voice does nothing to enhance the message that is already available to all in written form.

The memorized speech is the manuscript committed to memory and delivered without actually holding the paper. Most of the same problems facing the speaker who uses a manuscript are present for the giver of the memorized speech. If you are a talented actor, capable of delivering words prepared ahead of time as if they were the inspiration of the moment, this choice of delivery might be for you. We have to tell you that such talent is rare.

Most memorized speakers sound just that way: memorized. They display little willingness to adapt to feedback, little feeling of communicativeness, and little sense of sincerity. Because these factors are such a serious source of entropy, many speech teachers do not permit either the manuscript or the memorized speech. They are rarely heard in the usual business setting, for listeners in proposal and review meetings tend to interrupt the speaker. It makes little sense to prepare a definite set of sentences when you know that someone will probably make a comment that forces you in a direction you did not intend.

The extempore speech is a carefully prepared set of ideas and documentation, but the actual wording is formulated at the time of delivery. You have already been taught how to research and organize a presentation, so the foundation for the extempore speech is in place. We advocate this choice of presentation. It seems to incorporate most of the advantages needed for a successful presentation, while holding down the more serious disadvantages.

The extempore speech is never "off the cuff." Quite the opposite. The speaker will have prepared and practiced what will be said. Notes are almost always used, but those notes are usually an outline of the main ideas and the sequence in which those ideas will be presented. The notecards remind the speaker of the points to be covered, but the actual wording of the sentences evolves during the speech itself. As we said in Chapter 9, when transparencies are used, the visual aids can themselves serve as the notes.

Adjustments in wording can be made based on the audience's feedback—an obvious advantage to this method. Extempore speakers are much more willing to change the scope or order of what they say when they detect encouraging feedback than are those who have a definite set of words they are determined to speak.

The ability to look at and adjust to feedback also allows the speaker to concentrate on how the communication is being received. Watching an audience in order to correct the next set of messages encourages direct communication. As a result, extempore speakers do not display the formalized and stiff recitation quality that manuscript or memorized speakers sometimes show.

The disadvantage of the extempore speech is the lack of security a speaker feels without a prepared speech. A beginning speaker often feels threatened by the need to speak *with* an audience in the same way one speaks with friends. It seems preferable to read or recite *at* people than to form words that meet their needs.

Coming spontaneously as it does in an extempore speech, the wording often does not have the formality and polish some expect from public communication. The extempore speech tends to be quite conversational, so if you expect "oratory," you will be disappointed.

Some excellent extempore speakers overcome some of the disadvantages we have discussed by completely preparing certain portions of the presentation. The language in these portions might be very polished and eloquent. When inserted into a conversational flow of ideas, the result is quite effective.

As beginning speakers, you might benefit from preparing the exact wording of both the introduction and conclusion of your presentations. The security of knowing the first few sentences you will say helps you get through those first few moments of panic. A precisely worded conclusion gives a sense of eloquence as the speech ends.

With a carefully prepared speaking outline, some precisely worded portions, and the willingness to engage in direct communication with an audience, you have the makings of the most successful presentation choice.

The impromptu speech is one that has been unprepared for that moment. It is "off the cuff" and "I didn't know I was to say anything" speaking. Some people argue that it is very difficult to imagine a truly impromptu speech. Most of us have thought about and have strong opinions concerning many topics. Your presence at a meeting generally means that you know something about or are interested in the subject under consideration. In that sense, you have prepared. At a given moment, someone asks "What do you think about . . . ?" and you put into words what you have already concluded.

The impromptu speech demands a breadth of experience and knowledge. Calling upon a comprehensive background is not difficult for an experienced person. For the young or those lacking in information, the situation can be quite frightening. Few experiences are more humiliating than the realization that you can make only the most mundane comment because you know so little about a topic.

We do not believe that there is anything absolutely right or wrong about any of the choices of delivery. The manuscript speech has its place, and we have heard talented people deliver magnificent memorized speeches. The impromptu comment is sometimes necessary, and the extempore speech has its appropriate place. As with all communication variables, the moment and the people determine which kind is right. On balance, however, we advocate looking first to the extempore delivery; although not offering the security of some of the others, it does hold the potential for direct communication.

Visual and Audio Aids

In Chapter 9 we examined the variety of communication aids available. The technical professional must become familiar with most of them, for they form an integral part of the business world.

We have argued that certain kinds of aids are expected in many technical presentations. You are urged to review how you make and use the more common ones. One question remaining for public communication is, "How do you decide which (if any) to use?"

We offer four guidelines to follow in answering that question:

1. Does your listener expect some visual/audio aid?
2. Is your message enhanced by using an aid?
3. Is the room conducive to the aid you select?
4. Is the equipment available to prepare and use effectively the aid you select?

If you can answer "yes" to these questions, an aid is highly advisable.

ADDRESSING THE PROFESSIONAL AUDIENCE

The technical professional will be called upon for public communication in at least four specific places. One executive of a large international corporation told us that, if a person is not invited to participate in these settings, indications are that the career is stalled. We expect your career to build, so we expect to find you on the public platform of professional organizations, sales/contract presentations, lay community meetings, and the television camera.

Professional Organizations

Everyone we interviewed for the background of this book told us that their corporations, both in the United States and in Europe, expect participation in professional organizations. A young person's first introduction to the others in the discipline is often through the medium of a yearly professional society meeting. This person will eventually submit a paper to the organization and have it chosen for the program. Delivering the speech to the organization is the topic of this section.

Convention papers usually result from work done while fulfilling a contract with a public institution or agency. Employees must seek permission of their corporations to submit work results. Management decides if the work is such that a public statement ought to be made. Corporations are usually very careful about publicizing results that their competitors might use. An employee might be told that only certain portions of the work can be discussed, while others should not be mentioned at all. In addition, the sponsoring agency must give its approval in order to protect contractual interests.

In the academic world, publication of one's full research and findings is encouraged. On the other hand, the competitive business world sometimes makes the most interesting results a matter of secrecy. Just as often, when permission for publication of a portion of a work is granted, certain areas or questions are not opened to comment. We found widely recognized the realization that much of a professional's work might not be available for public discussion because of company restrictions.[5]

If your corporation grants permission for submission of a paper (or some portion of it) to a professional organization, it will be considered by the review board of the organization, which decides what papers warrant presentation. If your paper is selected, you will be invited to appear on the program and present the paper's findings during a specified period of time. The time allotted is usually far less than the time required for the entire paper. Your first task is to review your work and select those portions that can be discussed in the time available. You should have copies of the full paper, with contractually sensitive areas deleted, available for any audience member who wishes to have the full results.

It is neither necessary nor desirable to *read* the paper when the time for delivery arrives, although many people choose this approach. Some of the most effective presentations we have seen at technical conferences were those where people spoke from notes extemporaneously, helped by well-prepared visuals. We recall one conference where several speakers read from manuscripts while standing behind a tall podium. One perceptive man asked the person introducing him for a microphone he could secure around his neck. He then left the podium and delivered his paper from notes in a direct conversational style while he stood at the very edge of the raised platform. The contrast between the droning recitation of the others and this man's directness was phenomenal. Audience members talked of little else at the social gathering that followed.

This story brings up at least two things a young professional must contend with during the presentation: the introducer and the microphone. They warrant consideration.

The Introducer Whoever is chosen to acquaint the audience with the next speaker is extremely important. This person sets the stage for the speaker's credibility. Basically, the introduction is designed to convince the audience that the next speaker has something substantive to say. Most introducers cite the speaker's past work, honors, background, and reasons for being on the platform at that particular time. Speakers can help their introducers immensely by providing them with background information well ahead of time. A résumé or some abbreviated version of it is usually sufficient. If you wish specific information to be included (or excluded), contact the introducer. Most will welcome suggestions.

Microphones Amplification devices are tremendous machine aids. They have progressed from large microphones that seriously distorted sound to highly sensitive instruments small enough to be clipped inconspicuously to a lapel. Nevertheless, many people use them badly. A few suggestions are in order, for you will almost certainly need a microphone during a conference presentation.

Make certain that you know the kind of microphone you will have. If it is stationary and secured to a podium or the floor, you will be tied to it, unable to move away. Be careful to direct your words into the range of the microphone's capability. If you step away or to the side, unpleasant voice fading results. Allow nothing to get between the microphone and your mouth. If you are using a manuscript, hold it on the far side of the microphone, so the papers do not block the air force of your words. Do not touch the microphone or allow anything to brush up against it. The sound will carry with surprising and distracting force.

You might have the option of a movable microphone. Called *lavaliers*, these microphones are small devices that you loop around your neck by means of a cord. They fasten in front and rest on your chest. Their advantage is that they allow you to move freely. You take your amplification system with you so to speak. Their disadvantage is that they are hooked into the power source by means of a long thick cord. This cord has proved disastrous for unsuspecting speakers who trip over it. A good habit is to move the cord behind you, tucking a part of it into your waistband or belt. This holds it in place and out of the way of your feet.

A third type of microphone is a cordless transmitting one. Its system contains a radio transmitter that broadcasts the signal to a nearby receiver that subsequently amplifies the signal through the audio system. This microphone allows complete freedom of movement. Unfortunately, it is not yet widely available.

Whether stationary or mobile, modern microphones are quite sensitive, so speak into them in a natural tone of voice. Contrary to the practice of many rock singers, it is not necessary to "eat the mike" by moving very close to it. Speaking in a normal tone of voice from a distance of about eight inches should be adequate. If your usual speaking volume is very soft or very loud, you will have to adjust for the right sound. Very loud voices boom when amplified, while soft ones still cannot be heard.

The professional organization is a unique and promising place for technical specialists to enhance their reputation. Your corporation will probably encourage participation. We hope that your platform competence will boost your career within or beyond your home company.

Sales/Contract Opportunities

We have focused on the way corporations use their technical people to sell products and ideas because our experience tells us how important these activities are. We would like to make certain you are clear about the way proposals work and the peculiar hybrid public-speech model that has evolved in this setting.

The Proposal as a Sales Campaign In Chapter 6 we discussed the steps involved in going from proposal to the contract. As with any campaign, a proposal involves a variety of people and types of messages designed to achieve a goal. The goal of a proposal is to have the customer sign a contract for the activity the proposal specifies.

Thinking of the proposal as a means of selling an idea to someone else allows the technical staff to give proper attention to the customer as receiver of a message. One executive we interviewed told us how difficult it is to convince young technical people that ideas have to be sold before they can be developed. "Good ideas are plentiful. Ones that convince a client to invest in development are rare. The job of my staff is to sell ideas so they become real."

When people accept such a view, they are likely to begin thinking in terms of the receiver. The development of a product, in large measure, depends on how worthwhile someone thinks that product is going to be. We are aware of one large American corporation that scrapped development of a mechanism that had potential for a multimillion-dollar-a-year profit. The decision-makers could not be convinced to invest the funds needed to bring the product to production. Nobody criticized its scientific documentation. It was acknowledged to be an excellent and well-researched innovation. Its technical developers, however, failed to run the sort of sales campaign that got the product accepted by its own upper management. Complete with patents, the product remains only an idea on paper. Such an example is repeated often in corporate life. Scientists have difficulty recognizing that the selling of an idea might be more important than the scientific worth of the idea itself—at least in the corporate setting.

The Hybrid Public Speech The blend of public and group speaking that business uses is important to selling, defending, and publicizing corporate ideas. We have told you how in-house and customer reviews, as well as proposals, use a combination of group meeting and public speaking. We have named this combination the "hybrid public speech." A professional prepares to present a portion of a report to the assembled group. This speaker carries prime responsibility as sender of that portion of the message, and so speaks for an extended period of time. The group members are free to interrupt, question, comment, or add information at will. The group, rather than the individual speaker, has responsibility over the product of the speech. In this sense, the speech is a cross between the public and the group model. Because it is common, you need to consider some ways of preparing for it.

The choice of delivery almost certainly has to be extempore. A manuscript or memorized ten-minute presentation would seldom be allowed to continue as planned. Audience feedback is high, and the need to alter ideas based on interchanges makes specific wording unlikely. If the speaker could not adapt the remainder of what she or he had planned to say to the wishes of the questioner, credibility would be lost. Imagine the impact on a customer when the giver of a memorized speech says, "Sorry, I do not plan to talk about that. I have my next sentences committed to memory, so you have to listen to them."

Corporate hybrid public speeches are carefully prepared and usually extempore. The sequence of ideas and responsibility for presentation are assigned to each speaker well ahead of time. The organizational pattern of each speech is reviewed, as are decisions about how each speaker will build for the following speaker. At crucial stages, a "dry run" allows members of the team to hear how the whole package goes together. As we told you in Chapter 7, the process has no shortcuts. The pattern of research, organization, and practice repeats itself constantly.

The technical problem-to-solution method of organization is probably the one chosen. You might be responsible for any single portion or combination of portions. You would also be responsible for preparation of the accompanying visual/audio aids. The most frequent pattern is to use transparencies to lead the listener through the presentation. It is difficult for the one of us who works in the corporate setting to recall a single in-house or customer review in the last five years that did not use transparencies.

The speaker must be familiar enough with the subject to change the order of the presentation (and the subsequent transparencies) if a listener's comment indicates that some other sequence must be followed. Technical speakers become extremely familiar with the need to think quickly on their feet. Various exercises are tried during the dry run to give the participants experience with the "what if . . . ?" interruption. They try to come up with ideas about answering a question or framing an alternative pattern if some comment or objection is raised. Because interruptions will surely come, the memorized or manuscript speech is just as inappropriate as the impromptu would be.

In a very real sense, the business of technical professionals is to sell ideas and products so that they can be developed. Audience analysis, research, and proof are very much a part of that process. Persuasion is inherent to the job that must be done.

ADDRESSING THE GENERAL PUBLIC AUDIENCE

Technical specialists serve as important sources of information and criticism to people not trained in the scientific disciplines. Part of any educated person's job should be the sharing of knowledge with the wider community. Modern business practice encourages such reaching out because, among other things, it is an excellent source of good public relations. Between the duty of community education and the need to enhance a company's image, technical professionals quite often find themselves on platforms before lay audiences. We ought to look at three ways these contacts involve the technical speaker.

1. The Technical Person Serves as a Translator of Technical Language In a real sense, communities benefit from those members who can take specific theory and explain it in an understandable way. Schools bring in biologists or chemists to explain to children how life cycles work or how food additives affect the body. Universities invite physicists to enhance seminars by explaining new developments in energy conservation. Townships seek out engineers to explain how power stations work. Study groups, service groups, and special-interest organizations look for computer specialists willing to demystify the idea of programming. These professionals are in positions to spread accurate information, and many do so with a sense of duty.

When a trained mathematician prepares to translate the highly complex theory of celestial mechanics into a speech that explains how planetary motion is plotted for space shots, that professional faces a number of problems. The audience analysis is necessary to reveal what level of language, what use of jargon, what use of examples, and what visuals will be necessary. Aimed incorrectly, the speech will either insult by being patronizing with oversimplifications or confuse with its complexity. Adjusted to the particular audience, the technical person can make a real contribution to the pool of community knowledge.

2. The Technical Professional Should Be a Critic of Popular Information Sometimes the service a professional can perform involves correction or criticism of existing knowledge. The popular press often publishes highly inflated or only half-accurate data about a current scientific topic. For example, during the wave of popularity for solar home-heating, articles claiming that solar energy provided a nearly free heat source appeared regularly in magazines. Engineers working with solar power knew the data were inaccurate. The energy system could save a respectable sum of money, but the claim that "I never paid a dime for heat since I installed solar collectors" was simply untrue. Many engineers spoke to groups to correct misconceptions. The benefits to a community are obvious. Who can provide solid information and correct errors better than those trained specifically in the subject matter?

3. The Technical Professional Can Be a Public Relations Resource Corporations find that encouraging their employees to seek out the community platform is good business. When an astronomer is introduced to the local Rotary Club, the astronomer's company or institution gets good advertising. If the astronomer delivers a well-received presentation, the name of the firm is enhanced. Because this kind of contact improves relations with the public, some companies organize a speakers' bureau.[6] Even if there is no formal speakers' bureau, most companies recognize the benefit of having a competent representative carry their name to the wider community via the public platform.

The desire to achieve public recognition, however, is no excuse to divulge privileged information. Because of restrictions on classified work or the secrecy that sometimes surrounds certain product development, you have a responsibility to clear your presentation with your supervisor. Data accumulated during working hours are seen as the property of the employer, and serious repercussions result if you discuss that data publicly without permission. Even sensitive subjects can sometimes be cleared if the speaker agrees not to divulge certain facts. However, topics developed on your own time, such as personal hobbies or crafts, are your property to do with as you like.

With sufficient safeguards to make certain that no policy is being violated, going out into the community as your company's spokesperson is often rewarding. Not only do you gain a platform from which to enhance your reputation, but you also gain experience, and your company gets a good plug.

ADDRESSING THE MASS MEDIA AUDIENCE

The term *mass media* includes the many traditional channels of communication such as print, film, broadcasting, and wired communication. It also includes some innovations that Frederick Williams has called the "new technologies."[7] These include digital radio and television, 3-D television, and even a printed page that is "heard" by means of a scanning device that converts print into sound. Williams points out that newer media never completely replace the older ones, but they do alter them and tend to be used for different purposes.[8] In Chapter 9, we pointed to the prevalence of various communication aids in the business world. If Williams is correct, these will expand in purpose, and innovations will alter the ones currently used. Many of these aids fall under the definition of mass media. A competent technical professional will have to keep current with their development.

Television

Perhaps television is the most pervasive mass medium in the modern world. It brings special opportunities and problems for the technical professional. In the mid-1970s Hiebert, Ungurait, and Bohn stated that television is the costliest of all electric media because it must schedule material for eighteen hours per day.[9] That estimate is conservative in light of modern twenty-four-hour scheduling.

The business world was quick to see how its personnel might be called upon to use television, and companies moved to train people for competent appearances before the camera. Raymond Ross points out that General Motors has 680 people taking television-appearance courses. Oil companies, public utilities, and commercial firms train people to look and sound good on television. Most startling is Ross's statement that "It has been estimated that business and nonprofit organizations produce more television programming for in-house use than was carried by all four national TV networks."[10] Such trends show no indication of diminishing, and bright upcoming professionals will probably find themselves facing a camera eventually. Consider the possibility.

Talk shows are always looking for opinions from experts in various disciplines. News shows like to include statements about topics from those who supposedly have special knowledge. Documentaries, both in-house and commercially produced, often rely upon the commentary of experts. Members of companies with a newsworthy product or happening are routinely interviewed for inclusion on the evening (or morning, or midday, or late evening) news. The expertise demanded by those who want to understand physical and scientific occurrences is the sort the technical specialist can provide. Who is in better position, after all, to comment on the stresses that might have caused the collapse of a bridge, the interpretation of tornado warning systems, or the accountability of a firm that decided not to provide safeguards for its dangerous waste products?

Technical experts are called upon to explain and render opinions about many subjects that the lay audience wants to understand. The probability of your standing before a television camera in some professional capacity is quite high. You need to prepare for it.

Television commentary can take place on-site or in the studio. There are some significant differences between the two.

The on-site television taping can occur because some event just happened and an expert is invited to explain it. If you work for a consulting engineering firm investigating a failure mechanism for a collapsed bridge, it might seem a good programming device to stand you near the structure and tape your explanation of why the support fatigued. Present at such "shoots" are people to help you through the taping. The on-camera newsperson usually introduces you and plants some leading questions for your response. This interviewer will probably hold a hand-held microphone and can be counted on to direct your speaking into it. The words are being recorded and can be altered for later broadcast. At least one camera person will tape the visuals. The camera will be a portable model, most usually supported on the shoulder.

On-site reporters routinely deal with production under distracting and difficult conditions. They generally are quite proficient at securing good results. One method used to insure usable footage for later broadcast is taping far more than will be used. You might be surprised to find that your half hour of explanation was boiled down to a twenty-second clip. Realize that when the crew returns to the studio they will edit the verbal and the visual to fit their time requirements.

In-studio taping appears to the novice to be much more polished than the on-site. Certainly there is larger equipment, a permanent set, and tight control over lighting and sound. Look at the photograph in Figure 13-1. It contains most of what you can expect to see in the television studio. In addition to the person who will introduce you and the camera operator(s), there is usually a floor manager. This person tells you where to stand or sit, when to begin, and when to

Figure 13-1 Photograph of a television studio. Photo by Kate O'Shea.

speed up or slow down. The floor manager is connected to the control room, as are the camera operators, by means of earphones. The show's director and other technical staff watch the program's development on monitors from the control room. They speak their directions to the camera operators and the floor manager who, in turn, signal the person being taped.

Few shows are aired live. Material is usually taped for editing and use at some later time. Whether live or not, however, the control room staff can adjust lighting, sound, and camera shots while the action is happening. The picture we see on television, for example, is one of several possible ones that the director selected. Camera number one might be trying a close-up while camera two is using a medial shot, and the director then decides that the close-up does not look good enough for live or recorded transmission. With the delay in what is shot and what is sent, the director can alter any action as it develops.

Studio tapings are sometimes more frightening than on-site shoots because the crew has time to take light-meter readings, fool with the set, position the cameras, and otherwise alter the event. Television studios are filled with activity, much of it extremely frightening to the novice. At the same time, studio personnel can generally help the guest through the performance. The interviewer will probably chat with you to help you relax. The floor manager will explain the signals used during the taping. The camera operators will do their multiple adjustments before anything begins, so that you get used to their movement. When the moment comes for you to speak, however, keep a few suggestions in mind:

- Wait for the floor manager's signal.
- The operative camera has a red lighting signal, and you direct your eye contact there.
- Visualize a friendly face in the operative camera's lens. Talk to that friendly face as if a person were there before you.
- Do not worry too much about sound levels. Sensitive microphones pick up your voice, and the people in the control room will adjust the volume.
- Do not, however, neglect clear and well-paced delivery. Nothing that the control room crew can do will make a muddled or badly paced voice any better.
- Do not forget that receivers are going to decode your message. Speak to them so that they can decode the message as you wish.

Remember that the constants of communication are still there: you the sender select language systems to send a message to a receiver who re-creates it. You may never get the feedback, but human beings still react to what you say and do. Because you bear prime responsibility for the sending of the message and do so for an extended period of time, the television medium is the modern equivalent of the public platform. It is a potent communication channel, and you may find yourself using it.

WRAP-UP

We have used the term *public communication* in the sense that a presentation is public when a single person has prime responsibility for sending the message to more than two other persons who, in turn, yield their sender role and remain fairly silent until the speaker has finished. In this sense, technical people use the public speech frequently in their professional and private lives. When a parent lectures four children and expects that they will listen quietly, when a teacher presents the morning lesson to a class of thirty, when a geologist summarizes test findings to the project team, or when a biologist addresses a dinner meeting of the Lion's Club to explain the basic structure of DNA, these people are engaging in a public communication.

Many public settings are fairly formal, and so the speaker has the opportunity to specifically focus the variables of communication. As a result, the public setting forces a speaker to recognize the goal of the speech, the makeup of a particular audience, the verbal and nonverbal components of the message, the choices of delivery, and the selection of aids.

Professional life contains many opportunities for public presentations. Technical specialists who develop a successful career will find themselves in at least four public settings.

First, membership in professional organizations, highly encouraged in the business world, increases the possibility for a young person to present a paper at the organization's meeting. This splendid opportunity to gain a personal reputation is enhanced by skillful use of the microphone and visual aids.

At the heart of the business world is the marketing of some idea or product. Technical professionals become involved in the process of persuasion, which usually takes place as part of a campaign directed at a proposal or contract customer. The technical staff is expected to present ideas, defend against criticism, and deal with suggestions. The setting usually requires the hybrid public speech mode.

This hybrid speech is the peculiar adaptation the corporate world has made to carry out its business. It consists of a series of public speeches by specialists involved in some segment of the proposal or contract process. These people have prime responsibility for presenting and defending the material within their expertise. Their listeners consist of customer representatives, in-house personnel, and sometimes people from other corporations. The listeners, however, are fellow problem-solvers, and they expect to interrupt with questions and comments, for it is this group that has responsibility for the consequences of the speech.

The community outside the corporation also calls upon the technical professional for shared knowledge, and the public platform permits such sharing. A technical specialist can translate highly complex information into understandable language. The specialist also can correct and criticize popularly held conclusions about technical subjects. Finally, he or she acts as a public relations representative for the parent company. Because these functions are so important to the individual and the corporation, speakers' bureaus and public presentations are generally encouraged.

The machine age has expanded mass media avenues, and every indication is that the expansion will continue. Television, the most pervasive medium, makes it likely that bright ambitious persons with technical expertise will find themselves commenting before the camera.

For the educated citizen of the modern world, the public platform is more available now than it ever was in history. William D. Brooks says that this opportunity permits society to fulfill its public work. "There has never been more public speaking in the world than there is today."[11]

ACTIVITIES

1. Prepare an eight-to-ten-minute presentation to the class. Your eventual goal is to activate members to do some specific action. In preparation for this presentation:

   - Formalize the goal into a carefully worded statement.
   - Do an attitudinal and demographic audience analysis.
   - Isolate any potential nonverbal message problems.
   - Identify any jargon difficulties or other verbal language problems.
   - Choose the manuscript, impromptu, or extempore speech. Justify your selection of delivery.
   - Select your visual and/or audio aids.

2. Role-play a professional intercompany meeting in class. Several students assume the roles of people who are responsible for presenting their specific company's position on some portion of a contract. The remainder of the class should act as customer representatives or other-company representatives. Their role is to enter into the presentation with questions, objections, support materials, or elaborations. The speakers should adapt their presentations to this hybrid public-speech situation.

3. Each person in class will draw a topic from a list provided by the instructor. With only a few minutes' thinking preparation, present a fairly brief two-to-three-minute opinion speech on the topic. Make every attempt to make the speech meet your goal of motivation, action, and ego satisfaction.
4. Repeat the presentation described in Activity 3, except this time the speech is videotaped. On playback, evaluate the speech you have just seen yourself presenting. Pay special attention to your nonverbals as they were revealed on camera.
5. Attend one of the presentations by guest lecturers who come to your campus. A famous or noteworthy person would be best. Evaluate the presentation, using the evaluation form your communication instructor uses on your class members when you present your public communications. How does the guest lecturer fare on your class evaluation?

NOTES

1. Gary T. Hunt, *Public Speaking* (Englewood Cliffs, N.J.: Prentice-Hall, 1981), 4–5.
2. John F. Wilson and Carroll C. Arnold, *Public Speaking as a Liberal Art*, 5th ed. (Boston: Allyn and Bacon, 1983), 4.
3. Wilson and Arnold, *Public Speaking*, 4–5.
4. Fredrick Trautmann, "Effect: False Friend In Rhetorical Criticism," *The Pennsylvania Speech Annual* XXV (September 1968), 10–18.
5. Among those who commented in interviews on this restrictive policy were Lars Wilhelmsson, Volvo Flygmotor, New York City, March 6, 1984; Ulf Hall, Volvo Flygmotor, Trollhattan, Sweden, May 16, 1984; Frederick A. Chapman, Mercedes-Benz of North America, Montvale, New Jersey, April 2, 1984; Gunter Alisch, Daimler-Benz, Sindelgingen, West Germany, May 18, 1984; and Hans Juntgen, Bergbau-Forschung, Essen, West Germany, May 29, 1984. The policy is also widely applied in General Electric Corporation.
6. General Electric, Valley Forge, for example, has a well-organized and fully functioning speakers' bureau. Persons volunteering to appear before groups are listed in brochures, helped in preparation of their topics, and honored in a yearly luncheon. If the request for a speaker requires absence from work for a portion of the day, permission is usually granted. The cost of such activity is obviously seen as being worth the good public relations it buys.
7. Frederick Williams, *The New Communications* (Belmont, Calif.: Wadsworth, 1984), 160–74.
8. Williams, *The New Communications*, 162.

9. Ray Eldon Hiebert, Donald F. Ungurait, and Thomas W. Bohn, *Mass Media: An Introduction to Modern Communication* (New York: David McKay, 1974), 304.

10. Raymond Ross, *Speech Communication Fundamentals and Practice*, 6th ed. (Englewood Cliffs, N.J.: Prentice-Hall, 1983), 321.

11. William D. Brooks, *Speech Communication*, 3rd ed. (Dubuque, Iowa: Wm. C. Brown, 1978), 257.

CHAPTER FOURTEEN

EXPERT TESTIMONY

Communities rely on their technically trained people in at least one highly specialized way: for use as expert witnesses. The skills involved in becoming a good witness are very similar to those needed whenever good and clear opinions must be expressed and then defended. In this chapter we examine the nature of expertise and how experts help to determine policy. Furthermore, we suggest that the skills needed for this essential responsibility are the same as those required in daily communications where expertise is called into question.

You might act as a witness in a number of ways, the first so casual that it might not seem a question of expertise at all. Perhaps your neighbor stops you and says, "Al, you're a biologist. Do you think the construction of that office complex down the street is going to raise the bacteria count in our well water?" The neighbor listens to your opinion and perhaps even accepts it because of your expertise.

Suppose, at some later time, those who own that office complex ask your township zoning board for a variance to allow expansion of the facility. You might be called upon to witness in a second type of setting called a hearing. A hearing can take place at all levels: local, state, national, and international. It is a meeting where those who must decide policy can hear the opinions of people having various viewpoints. Input is sought to help the decision-makers arrive at a policy.

As a biologist, especially if you have credibility in your community, you might decide (or be urged by others) to appear at the zoning hearing. Speaking as an expert in microbiology, your opinions about the effects of further office expansion on the safety of surrounding well water carry far more weight than would the opinion of some other nontechnical professional.

Finally, suppose the zoning board decides to grant the request for expansion, despite all the neighbors' efforts. In a year or two, the community's worst fears are realized: the well water becomes polluted, and many people become ill. One of them decides to sue the complex managers, and you are hired by the attorneys to testify in a court of law. This third use of a technically trained person as an expert witness is not at all uncommon. Francis Wellman, in his landmark book on cross-examination, says,

In our American courts, as they are now constituted, I think I am safe in saying that in half the cases presented to a jury the evidence of one or more expert witnesses becomes a very important factor in a juror's effort to arrive at a just verdict.[1]

What is startling about the statement is that Wellman wrote it in 1903. Imagine what the estimate would be in today's "Let's sue" atmosphere with its reliance on technology to measure fact and establish proof.

From the casual exchange of opinion among peers, to the hearing where personal conclusions are formally submitted to help structure policy decisions, to the law court where conflicts are judged, the technical professional serves as a unique resource. We need to examine the nature of expertise and help you prepare to render your opinions competently.

EXPERTS AND SOURCES OF COMPETENCE

How expertise is determined seems obvious at first. An expert ought to be someone with special knowledge, yet it does not always happen that way. Advertisements show you that many popular figures speak in testimonials about a variety of products. A film star or a sports figure speaks in favor of everything from breakfast cereal to washing machines. We admire these people, perhaps consider them personal heroes, so we judge their words as more important than those of others. The film star might also support a public policy, and the sports figure can campaign for a given politician. Such actions are within their rights, but their opinions about policy, candidates, or washing machines do not derive from expertise.

We shall consider people *experts* when their background includes one or more of the three competence sources we list.

Education

People can claim expertise when they have a recognized level of "book learning." We assume they know certain facts because they have been through the course-work that teaches those facts. The point where some level of learning is recognized as expertise is not a matter of a specific number of classes. Certainly, taking a single art history class would not make you an expert on Renaissance painters. If you take many art courses, most dealing with the Renaissance, however, you finally attain recognition as "one who knows." People weigh your opinions about da Vinci and Michelangelo differently from those of your economics-major roommate.

Technical people, like the Renaissance expert, derive their expertise from the course-work they have had in some special discipline. We assume that those who hold doctorates from established institutions know more than those who have a master's degree or those who earned their doctorates at unrecognized schools. The newly hired employee with a bachelor's degree is assumed to be more knowledgeable than the freshman, but less so than the supervisor with an advanced degree. We make assumptions about the knowledge people have, based on how many courses they have taken, even though the assumption might be dead wrong. We simply give credibility to those educated in some body of information.

Experience

Another source of expertise is the training achieved through actual hands-on accomplishment. Education and experience often go hand in hand, but the new college graduate is assumed to have less expertise than someone who has on-the-job experience in addition to a degree. This viewpoint is a source of persistent frustration to the newly graduated job hunter.

Stories abound about people who learned complex skills but had no formal education. We assume that doing something time and time again gives the doer special knowledge about the task. When we spoke in Chapter 11 about "quality circles," we were describing the business approach to knowledge by experience. Quality circles, you will recall, bring together people who do certain jobs every day. These people, many with no theoretical education at all, are assumed to know how their jobs can be made efficient and their products improved. Their opinions are respected by management because they have experience in making that product and working at that job.

Physical Position

The third source of expertise derives from a person's being somewhere at a crucial time. Observers of bank robberies or murders are called to be witnesses because they happened to be there when the crime happened. People who were at the site of a catastrophe are in a unique position to tell what happened. We listen to such testimony seriously, for often the witness who was physically present is the only source of accurate information.

We also assign an expertise to those whose jobs give them access to certain places closed off to the rest of us. If only one person is permitted to enter the test chamber, we rely on that person for information about conditions there. If only vice presidents attend certain meetings, we assume that their reports about those meetings are accurate. We rely on them because they have knowledge we cannot have. Such people, whether witness to a hijacking or the company president's tirade, can give us reports about "what really happened." As such, they are expert.

Testing for Expertise

Because we rely on witnesses to tell us what we cannot know for ourselves, they have a unique responsibility. Our law courts and decision-making bodies listen to experts. The words of technical people can structure the way policy is formed. We need to have ways to verify claims of expertise to make certain that the testimony we hear is worth our consideration.

Any time someone asks to be accepted as an expert, ask yourself the following questions:

1. Is this person's educational background adequate for the conclusions to be drawn? As a source of special knowledge, exposure to book learning and course-work varies. The knowledge of a marginal student from Podunk University might be far inferior to that of a Superlative Institute honor graduate. Make certain your expert has the learning necessary for credibility. Ask: Is this person's educational background strong enough to provide superior knowledge about the point in question?
2. Is this person's experience substantial enough for the kind of conclusions to be drawn? Those who must make decisions concerning expertise must be convinced that the experience brings

with it unique skills. Ask: Has this person done this task enough to verify the special basis of knowledge claimed?

3. Is there a physical/situational constraint that prohibits the special knowledge claimed? An eyewitness who normally wears glasses must have been wearing them to claim to have seen an occurrence. A group member must have been in the room at the time the event took place in order to testify about the meeting. Ask: Did anything get in the way of this person's perceiving what is claimed?

4. Is the person speaking only in the specific area of personal expertise? Experts in one topic may offer comments on a variety of unrelated topics. The sports figure who sells washing machines might legitimately speak about the relative merit of sporting equipment or how comfortable different clothing is during play, but any claims of extraordinary experience with appliances should be questioned. Ask: Does this person's opinion relate directly to the education, experience, or physical position that would constitute expertise?

5. Is the person's reputation for personal credibility sound? Any doubt about the integrity of the individual will call expertise into question. People who lie, who steal, who distort numbers, who dishonor contracts, and so on place their entire reputation on the line. Most of us are unwilling to accept the words of a known charlatan. Ask: Is this person worthy of belief?

6. Has this person's previous testimony been reliable? Some individuals do not lie, but they do seem to make many mistakes in interpretation or prediction. We tend to rely less and less on someone whose past predictions proved false. Ask: Is this person's track record of reliability good enough to warrant acceptance of current conclusions?

7. Will the person's testimony result in substantial benefit to that individual? Receiving a reasonable fee for acting as a witness is an acceptable practice for experts who are hired because of their specialties. If, however, the expert is part owner of the company being supported by testimony, the expert's credibility is called into question. An eyewitness in a criminal proceeding who has been offered a reduced sentence for testifying is like-

wise suspect. Ask: Does this person gain enough as a result of this testimony to taint expert judgment?

8. Is this person relatively free of personal bias in regard to the issue under consideration? Sometimes an expert's opinion is colored, not for personal gain, but because of personal beliefs. If Jack believes that women make poor physicians, then his statements about the work of a female doctor have to be questioned. If Ginny thinks motorcycles are inherently dangerous, she would make a poor witness for a motorcycle-accident case. Ask: Does the expert exhibit any attitude that is likely to color perception?

9. Is the set of statements internally consistent? People often change their minds in the middle of statements. They begin by supporting and end up damning whatever is at issue. This erratic movement might be evidence that their thinking is extremely fuzzy. Looking to see whether the person's thoughts hang together consistently can reveal such a flaw. Ask: Does this person follow through with conclusions that mesh with his or her other conclusions?

10. Are this person's views consistent with those of other experts in the same discipline? Innovators in science are often regarded as heroes, but that acclaim is a long time in coming. Be wary of someone who claims something that most other experts deny. The expert might be a landmark thinker but might also be wrong. Methods, procedures, conclusions, or proof bases that fall outside the mainstream of a discipline must be handled cautiously. Ask: Do other experts in the discipline agree with this person?

The more the answers to these questions trouble you, the more likely you are to have a faulty expert. Remember that some decision-maker must *accept* this person as having special knowledge that is worthy of consideration. The tests hold for anyone claiming to be an expert. Whether the expertise is alleged to be about which car to purchase or which toxin could have caused the deformity in a newborn child, make certain the speaker is qualified. If you do not test your expert and present a faulty witness in a law court or hearing, be certain that the opposing attorney or hostile listener will do the testing for you.

EXPERTS IN LAW COURTS

Many have noted that ours is an era of the courts. People sue each other with alarming frequency, so the likelihood that any individual will be involved in litigation is increasing. Whether the willingness to go to court to solve conflicts is good or bad, the reality exists. At least two professions benefit from our rush to judgment: lawyers and technical professionals. Lawyers do so because citizens, even highly educated ones, cannot understand the intricacy of law. Technical professionals do so because lawyers and juries cannot interpret complex technology. A citizen involved in a lawsuit eventually has to seek out legal advice, and the attorney often has to hire an expert witness. Depending on your specialty in mathematics or science, some of you may be called upon to give testimony in a court of law.

Assumptions about Testimony

Testimony is a term used to designate the statements witnesses make about the specific areas of their special knowledge. In a criminal case, special knowledge (expertise) might be that of the eyewitness who claims to have seen the crime. Also offering relevant special knowledge might be the technical expert who presents such things as a ballistics report, an analysis of a blood sample, or an interpretation of a drug/alcohol level. In our sense, the eyewitness and the technical witness are both expert. Whether they are legitimately entitled to make their statements is a judgment. The supporting and opposing attorneys will both probe any witness's right to claimed expertise.

Most of us were raised on television or film courtroom drama. The brilliant lawyer who catches the expert in an error about blood samples, while defending a client against a murder charge, is a scriptwriter's dream. It happens in a real courtroom too, of course, but most trial work is not nearly so exciting. Trials are usually about accidents and who was at fault, "natural versus manufactured" flaws, fulfillment of contractual duties, and many other important but mundane human conflicts, rather than about sensational incidents. The eye-opening experience of observers who see a trial for the first time is the lack of glitter. Rieke and Sillars point out:

For many years practicing lawyers have objected to the image portrayed by Perry Mason, the fictional trial attorney. He cross-examines witnesses until they confess. Real lawyers suggest this simply never happens. On the contrary, only the most naive cross-examiner, after securing a series of factual admissions, goes on to ask the last question, "Then we really shouldn't believe what you said?"[2]

Despite our preconceived notions of what goes on in the courtroom, we need to recognize that the important issues being decided require careful procedures to insure a just decision. Such procedures are not always dramatic, but they do convince us that multiple rights are being safeguarded. The use of technical experts to render opinions, describe tests, and interpret findings is inherent to such a process.

Consider some of the assumptions that our legal system makes about testimony in general and technical-expert testimony in particular.

1. Testimony should be considered as a reasonable search for legitimate information. The controversies that law courts decide are complex. What *really* happened is extremely difficult to infer. Using the definition of fact and inference, much as we did in Chapter 3, the judicial system assumes that factual information must be monitored carefully. Experts recognized as competent can give interpretations and descriptions that allow more probability for correctness than do the statements of those with less competence.
2. Even the most factual testimony includes opinions. Patterson and Zarefsky point out that testimony is always secondary, "an opinion of reality as stated by another person or information recorded by another."[3] No matter how carefully an event is described, a test is documented, or a finding stated, those memories are interpreted by a living human. There is certainly a great difference between the testimonies that "The blood sample contained 0.5 percent alcohol" and "The brakes could not possibly have failed in the manner stated." In the first, the expert is factually reporting the results of a given test; in the second, the expert is giving a judgment. Yet, even the statement about the test contains implicit opinions. The expert maintains that the test (opinion) is valid, its procedures (opinion) were followed reliably, and its 0.5 percent reading (opinion) was recorded accurately. The expert presents the findings based on assumptions of exactness. Such opinions do not invalidate the testimony. Quite the opposite—the expert renders the statements as probability of exactness. Because of the credibility expertise brings, the decision-makers listen to the conclusions as words worthy of belief.

3. Testimony lends credibility to the ideas an attorney hopes to establish in proportion to the extent listeners believe the expert. To have impact, what a witness says must be accepted by the judgment-maker. Because expert testimony involves information of great complexity, usually beyond the knowledge of most people, the listener must accept the expert before accepting the expert's conclusions. If a listener questions the expert's character, methodology, or bias, then the expert's statements appear colored. Attorneys use testimony to bolster belief, showing that a person of special training is willing to state support of an idea shared by that attorney. Furthermore, the expert's description of how the conclusion is arrived at reinforces credibility. Jensen says of this highly subjective and judgmental process that

   securing expert allies does not guarantee that one's point is right or will be accepted, but it does lend strength to one's contention. The degree of strength depends on how acceptable the authorities are in the minds of the audience.[4]

4. Testimony is presented initially under the direction of the friendly or allied attorney and is known as *direct examination*, but it is also subject to the questioning of the opposition attorney and is known as *cross-examination*. The sequence of the statements made by a witness begins with a set of answers made in response to the allied attorney. This is the lawyer from the witness's side, those who have hired you to speak. When "your side" has finished, the opposition has the right to ask its own questions. In this presentation and examination, the soundness of the expert and the conclusions is thought to be insured. No evidence may be introduced unless those who found it (if it is an object), those who tested it (if it is a result), or those who examined it (if it is subject to interpretation), are available for testimony and cross-examination. Secondhand or hearsay testimony is not permitted. The witness who actually saw, tested, examined, found, or interpreted must be available for questioning on the witness stand.

5. Testimony and the cross-questioning process are assumed to establish probability of what *really* happened. The judges, juries, and lawmakers who listen to testimony were not present when the issues at hand happened. They are forced to rely on the reports of others. Underlying the judicial process is the assump-

tion that the slow and careful use of statements, subject to the inspection of an opponent, will yield a higher probability of accuracy than would other processes. Some legal theorists think that this right of confrontation has its origin in criminal proceedings and prohibits secret or ill-founded accusations.[5]

6. Testimony is assumed to be truthful. When a witness is presented as worthy of belief, the law enforces upon that witness standards of telling the truth. A person might recall something inaccurately or forget an action entirely, but the person is assumed to be speaking as truthfully as memory allows. Known distortions of facts, test results that were never recorded, or false identifications are subject to serious punishments. Freeley points out that this standard holds not only for a law court, but also for testimony before a government body. An oath is ordinarily administered, and its taker is subject to penalties of perjury or contempt.[6]

7. Testimony from different experts can and often does differ. The probability of establishing what really happened is just that: probability. Two equally trained experts may perform equally valid tests on an object and arrive at very similar data, yet their interpretation of what those data say can vary radically. Science cannot assume absolute certainty, and its practitioners cannot be held to such a standard. The best an expert can do is examine, run the required tests carefully, collect the resultant data, and render a judgment. If the judgment runs counter to what a colleague has concluded, it is the right of the decision-makers to choose between them.

8. Testimony occurs as an interaction between people and is subject to all the entropy losses of any other communication process. A law court or hearing chamber is such a formal and formidable place that it is easy to forget the basic humanity of the event. As a communication, it is subject to the sender, message, receiver, and feedback considerations of any other communication. Because of the nature of this particular exchange, an expert must remember that the message must be made as easy for the listener to decode as possible. The elements of good communication are as essential in a law court as they are in other settings. Because the stakes are so high, keeping entropy at its lowest level should be one of the expert witness's aims.

Preparation for Testimony

We believe that understanding what might happen in a court of law—or in any communication setting—will help you to prepare with competence. The invitation to testify will probably come from an attorney who feels that the client's case would be bolstered if you could run some tests and testify to the results of those tests. The attorney will help you prepare for what you will say and how the opposition will proceed. The following set of guidelines will help you in this preparation. It is similar to the preparation you would do whenever some opinion might be challenged.

Suppose that, in a motorcycle accident, a rider lost control of the bike and died in the subsequent collision with a truck. The rider's family decides to sue the motorcycle manufacturer, claiming that the bike's frame snapped under fatigue and caused the rider to lose control. The manufacturer says that the structure was sound, and the frame snapped only after the bike collided with the truck.

You are a metallurgical engineer and do upon occasion testify in cases where metallurgical questions arise. The family's attorney hires you to examine the motorcycle's metal frame, note the position of the break, run some stress and impact tests, and render an opinion about when the frame was likely to have snapped—before or during the impact. You make your observations and analyses, concluding that the metal showed evidence of failure due to fatigue and that the bracing structure was marginally designed. In your opinion, the frame could well have snapped before impact.

As you prepare for your day in court, keep the following directives in mind.

1. Review and prepare all pertinent documents, tests, dates, results, and methodologies. Memory is extremely unreliable. The backlog of cases on most court calendars makes it likely that a lengthy time will pass between the testing and your appearance in the witness chair. With the help of your allied attorney, review all materials. Anticipate questions from the opposition and prepare your answers. If your testimony involves long and detailed sets of numbers, have them on hand. Exactness of results is important.

The allied attorney will lead you through your testimony first. Answer as precisely as your testing permits. The acceptance of your conclusion by the judge or jury depends, in part, on how well they accept your expertise. Fumbling answers or erroneously remembered data do nothing to create a positive perception.

Although not required to do so, the opposition attorney will probably decide to cross-examine you. Careful preparation, accurate records, and reasonable steps leading to conclusion are your best defense against an assault on your credibility.

2. Prepare to establish yourself as an expert. The allied attorney will probably begin your direct examination by having you state your experience and training. This line of questioning is intended to establish your expertise and your right to render conclusions. As Tierney says, "The expert witness is one who is brought to the stand to do something that most ordinary witnesses are forbidden to do: give his opinion on some issue, rather than merely give the facts as he observed them."[7]

Be very explicit in what you say about yourself. Rather than stating "I am an engineer," say, "I hold a Ph.D. in engineering from Rensselaer Polytechnic Institute, and my concentration of courses and doctoral dissertation are on the topic of metallurgy and metal fatigue." Since your testimony is focused on whether or not you think the particular structure of the motorcycle's metal frame may have been faulty, the attorney will probably probe the nature of your expertise in this area. Have graduation and employment dates, honors, professional organizations, publications, and other personal-background materials ready.

Using many of the same tests for experts we have suggested, the opposition attorney may challenge your expertise by trying to establish that your training was insufficient to justify your conclusions, that your education did not deal specifically with the subject at hand, or perhaps that your method of testing the motorcycle frame is not a reliable one. Have your facts and information ready.

3. Discuss with your allied attorney any concerns about any portion of your expertise, your testing, or your conclusion. The witness stand is no place for surprises, especially if those surprises are the allied attorney's. If you have doubts, inform the attorney ahead of time. If there is something in your background that might discredit your testimony, make sure the lawyer knows it. The court knows you are being paid for your testimony, for example, and that is considered fair. If the trial reveals that the dead rider is your spouse's nephew, however, that would be seen as unacceptable bias.

4. Listen carefully to the questions of the allied and opposition attorneys. It is the duty of the lawyer who hired you to present you and your conclusions in the best possible light. Likewise, the duty of

the other side is to probe and find areas of doubt. In the final analysis, however, both sides seek the same answer: What really happened when that motorcycle and that truck collided?

In the spirit of openness, remembering that you are under oath, answer all legitimate questions fully and honestly. The witness stand is no place for mini-speeches, so watch carefully for feedback that tells you that your answer is complete. If you have more to say about the topic, allow the lawyer to probe for it. Resist the temptation to elaborate beyond the question.

5. Ask if you do not understand a question. Being an expert does not make you superhuman. Some attorneys ask their questions in an extremely confusing manner or state them in a way that is ambiguous. How would you respond to a query that says, "Do you not agree that it is not the duty of a company to guarantee safety under all conditions not predictable in design?" By the time you figure out all the "nots," you should realize that the question is confusing. Ask for clarification, such as "Are you saying that a company must design its product only for certain conditions and not for others?"

In the same spirit, admit the limits of your knowledge. If you do not know an answer, say so. Bluffing is only a shade away from perjury, and the risk is not warranted. If your pretense at knowledge is revealed, your credibility is ruined.

6. Know that you are required to answer all reasonable questions. Although you are entitled to be asked clear questions, you cannot hide behind a screen of "I don't understand." A witness who does not seem to know what the questions mean when they are clear to the rest of the listeners seems to be stalling or to be extremely dense. If you seem to be stalling, listeners suspect that you have something to hide. Inability to understand questions casts your entire intelligence into doubt. Neither evaluation helps your case.

Some witnesses choose to answer a legitimate question in very abstract terms, thereby avoiding a direct answer. In response to "Are you aware that the method you have chosen to test the motorcycle frame does not conform to the American Association of Metallurgists standards?" you say, "There are many people who continue to test as I have done." The answer is evasive. You can be sure the attorney will follow up with another question and drive home the possible unreliability of your methodology.

Refusal to answer is simply not acceptable. Besides the risk of facing a contempt-of-court charge, a refusal is often more damaging than any stated answer would be. The decision-maker is left with the impression that the lawyer had you in a telling point, and your only way of handling it was silence.

7. Insist that illegitimate questions be labeled as such and reworded into an answerable form. The ideal of questioning is to leave the respondent only a "yes" or "no" option. If you feel that the question cannot be answered in that way, but requires some qualifier, say so. If the opposition attorney says, "You really cannot say for certain that the metal broke before impact, can you?" note that the question deserves more than a simple "yes" or "no." You might try "There is nobody who can say for certain what happened besides perhaps, the dead rider, but I have stated my conclusion based on the best testing my experience and education permit. I am confident of my answer."

In the same way, some attorneys try the unfair tactic of the loaded question. These questions are similar to the "Do you still beat your wife?" type. Either answer will cause problems. If you face a loaded question, label it as such and ask for different wording. If the opposition lawyer says, for example, "Do you support irresponsible motorcycle operation by careless riders?" label that a loaded question. If you answer "yes," you appear ridiculous. If you say "no," you must defend the dead rider against the charge of irresponsibility and carelessness. Neither is appropriate. Say instead, "That is a loaded question. If the attorney will restate it so I can answer honestly within my expertise, I shall be glad to do so."

8. Practice courtesy to both the allied and opposition lawyers. The temptation to respond in kind to a loud and badgering style of questioning is great. As your expertise is challenged or your methods called unreliable, your temper begins to rise and a shouting match can follow. Resist the temptation. As a matter of fact, the more the lawyer provokes, the calmer and more reasonable you should appear. The contrast will be startling. Tierney says that

the best way to regulate his exchanges on cross-examination is by the exercise of self-restraint, and an absolute refusal to lose his temper, even in the face of provocation. . . . It is the regrettable technique of some counsel to try to provoke a witness so that he appears uncontrolled and vindictive to the jury, and this is a special reason for keeping a tight watch over what one says.[8]

The advice is good. When you feel anger rising, be on guard. The attorney probably wants you to be emotional. Rather than playing into the strategy, use the advice a debate coach gives to her students, "Sit tight. Speak right. Don't fight."

9. Practice good communication. As a communication setting, the court of law provides you with the opportunity to select language that will help the listeners decode with minimum loss of message. Be careful of jargon peculiar to your discipline. Even a trained decision-maker may not know the terminology you use. Provide adequate definitions and explanations for any complex symbols you use. Practice good sending (speech) and receiving (listening) behavior. Look especially for feedback that tells you how you are doing. If the looks indicate you are causing confusion, talking too long, or incurring hostility, use the feedback to alter your next message.

Remember what your goal is in the motorcycle accident case: helping a jury (or judge) to decide most probably what happened just before the rider lost control of the machine. Neither you nor the court can know for certain, but your expertise allows you to make some conclusions to help the court in its verdict. As an expert, you bring some significant support to your allied attorney. You are compensated for your aid financially, but the bigger reward is using your unique ability to provide a needed community service.

EXPERTS IN HEARINGS

The second formal setting where expert testimony is used is the hearing. A *hearing* is a fact-finding meeting where decision-makers gather a range of opinions from various viewpoints. This information is used to guide those in power to take some action.

Hearings have become extremely common at all levels of government. Local citizens are given the opportunity to express their opinions about changes proposed in their neighborhoods. If a property were under consideration as a toxic waste-dump site, the neighbors would probably have serious doubts about such a proposal. They might encourage those among them with chemical or biological expertise to express their opinions to the appropriate board. If no community member had such expertise, they might decide to hire someone to make their opposition to the plan known. This expert, often working with the group's attorneys, gives statements to the decision-makers in the form of testimony.

This pattern holds for state agencies, as well as national and international ones. Many legislative assemblies today have delegated a great deal of rule making to "regulatory agencies." Examples on the federal level are the Securities and Exchange Commission, the Food and Drug Administration, the Environmental Protection Agency, and the Federal Communications Commission.[9]

Much knowledge is required to enact the very special laws covering extremely specialized areas. Although the various agencies have considerable power to act, the complex nature of the issues they regulate requires technical knowledge beyond the training of agency staff or legislators. Experts are regularly invited or hired to present their opinions before these bodies. Pressure groups also hire experts so that their case is strengthened as they oppose or support pending legislation.

The kinds of issues and the people capable of giving information about them vary, but engineers and biologists seem to appear frequently. A sample might include the following cases.[10]

- Engineers to testify before the Subcommittee on Investigations and Oversight on resurfacing, restoration, and rehabilitation of roads.[11]
- Biologists from the National Institute of Environmental Health Sciences and several physicians representing the American Medical Association, testifying before a Subcommittee of the Committee on Veterans' Affairs on the Agent Orange Relief Act.[12]
- Registered Professional Engineer, Director of Public Services, testifying before a Subcommittee of the Committee on Public Works and Transportation on water-pollution control.[13]

In addition to these fairly typical agencies using the testimony of technical professionals, there are any number of subcommittees of bodies such as the Committee on Science and Technology of the House of Representatives; the Senate Committee on Commerce, Science, and Transportation; or the Committee on Public Works and Transportation. When these committees and their various subcommittees are added to similar committees and subcommittees of regulatory agencies, the need for expert information becomes obvious. This process is duplicated at the state and local level, as well as at the international. It should not be surprising, then, that technical people find themselves testifying far more frequently than they would have anticipated as students.

Similarities to Law Courts

In some ways the hearing is similar to the legal trial, but in others it differs significantly. Its similarities might be summarized by the statements that hearings and the law courts both:

- Rely on expert information as a necessary part of decision making.
- Expect the expert to speak from a legitimate basis of expertise.
- Require documentation of method and reasons leading to conclusions.
- Insure truthfulness, in many instances, by administration of oath.
- Expect the right to question the witness.

Admission of Biases

The expert appearing before a hearing is assumed to have strong bias. A court of law recognizes that an expert's opinions are never completely without biases, but the court also assumes these will be kept in check. Our metallurgist in the motorcycle example was hired to do certain tests and make certain comments by one side in the legal controversy. The metallurgist appears at the trial *because* of conclusions favorable to one point of view. Yet, the process assumes that the expert's background and methods allow those conclusions to be made within an acceptable range of objectivity. No such assumptions can be made about a hearing.

Those who hold hearings do so in order to judge viewpoints from various sides of some controversy. People can and often do ask to appear at such hearings. Organizations send their representatives to testify from the organization's point of view. The hearing directors might also *require* a person to appear before them. Such a demand for some person's presence is called a *subpoena*, and this legal document is one way of making certain that those who should testify actually do. Sometimes an organization or group of private citizens decides that they do not have the expertise to impress decision-makers at a crucial hearing. These people must rely on outside experts who, for a fee, agree to appear. Whether voluntary, subpoenaed, or paid, however, listeners assume that the expert will speak from a decidedly biased point of view. According to Tierney,

Almost certainly, a witness called before a regulatory agency will have strong views on the issue to be determined. It may be that this is the very reason for his testimony being solicited.[14]

As a person of expertise, speaking from a point of view that is admittedly in your own self-interest, you must recognize the means necessary for preparing and acting at a hearing. Witnesses are permitted (and even expected) to make statements before the hearing members. These statements are carefully prepared speeches of varying length whose aim is to convince the decision-makers that your side of the issue is the right one. These speeches are persuasion to action, because the speech is used to affect the way the decision-makers will do something. What we have said about the research, organization, and delivery of speeches can be used directly to structure the kind of statement you make at the hearing. In addition, you are permitted to submit any supporting documentation for the record. Copies of test results, log books, and other materials appear along with the statement you make. Together, these form your testimony.

Speaking about courtroom witnesses, Yarmey says, "Speakers never enter the courtroom as equals. Their credibility in the first instance is tied to the prestige people award them as representatives of groups or institutions."[15] The same is true of hearings. Because of the respect American culture has for science in general, scientists are regarded as people of prestigious status. The scientist who supports a certain position can lend a great deal of weight to that position, but an unprepared or incompetent-sounding scientist cannot. As a result, your expertise will be explored and evaluated by the listeners. Your statement will probably include personal information about yourself such as that supplied by one witness in an Agent Orange hearing:

Mr. Chairman and members of the committee, my name is Franz M. Enzinger, M.D. I am a pathologist and Chairman of the Department of Soft Tissue Pathology at the Armed Forces Institute of Pathology in Washington.[16]

Once expertise is established, the verbal and nonverbal behavior of the speaker either enhances or damages credibility. As in all communication settings, entropy at a hearing can increase to the point of distorting or destroying the message. Data and conclusions rarely seem convincing when the expert providing them appears to be fumbling and insecure (or arrogant and insulting). Keep that in mind as you prepare your statement, and let it guide you as you actually deliver it.

Finally, the hearing setting allows questioning by participants. If the subcommittee, for example, has nine members, any one or all of them have the right to question. Review the previous section on questioning in the law court. Most of the same cautions apply to the hearing, except that there is no "allied attorney" to lead you through your statement. You speak on your own and then the committee questions.

You have no guarantee that committee members will be friendly, and some might be hostile. Your viewpoint could be damaging to the "pet project" one member has. Anyone on the committee could react to your statement with an aggressive set of questions. Your position ought to be the same as that of the witness in a court of law. Do not allow yourself to be drawn into a shouting match. The more angry the committee member becomes, the more calmly reasonable you should appear.

Your best preparation for a hard line of questioning is security in your viewpoint. Assemble all your documentation, think through your methodology, and perform the fallacy tests on your conclusions. If certain members disagree with you, that is their right. The way you defend your ideas against the assault might just sway the judgment of other undecided members to your side.

In most instances, hearings are no more lively than the usual court trial. Assume that the one at which you speak is meeting to gather information for a decision on some future action. Give your opinions with the solid backing of your scientific training. As a concerned citizen, you have the right to speak; but, as an expert, you gain the respect that makes it likely you will be heard. Such advantage is considerable.

WRAP-UP

The person trained in science and its various applications occupies a very special place in our legal and legislative systems. Courts and law-making bodies rely on expertise for information. Competent technical professionals might well find themselves in the positions of expert witnesses.

Expertise derives from three sources: education, experience, and physical position at some crucial time. The first two are the most usual bases for the technical expert. People presenting themselves in such a role must be prepared to withstand the tests that determine whether expertise really exists. In the final analysis, an expert is one who is accepted by a listener as having the legitimate right to make the conclusions being stated.

The expert witness appears in two special settings: the court of law and the hearing. In each of these, the witness might testify voluntarily, through subpoena, or for payment. Unique among all the other witnesses, the expert is permitted to draw conclusions and is recognized as being paid for the statements made. A reasonable fee is an expected part of this process.

The law court involves testimony given in response to questions. The first set of questions, called direct examination, is asked by the attorney representing the side with which the expert is allied. The second set, called cross-examination, may (but does not have to) be asked by the opposition attorney. In either case, the expert witness must prepare for the questioning. We have suggested a number of guidelines and cautions.

The hearing is similar to a law court in some ways, but it is significantly different in others. Hearings are information-gathering settings, assembled to allow the airing of different viewpoints. These viewpoints are used by the hearing members to structure a future policy recommendation. As a result, people who appear at hearings are presumed to speak from a strong bias. They come to the hearing specifically because they have a definite point of view. Their testimony consists of (1) a speech of almost any length, whose goal is to sway the decision-makers to the speaker's side, (2) written documents for inclusion in the hearing record, and (3) responses to questions members might ask. The principles of good speech research, organization, and delivery must be used to enhance the credibility of the expert. In addition, the same cautions that apply to answering questions at a law court apply at a hearing.

The technically trained person is a unique community resource for expert testimony. When modern culture's respect for science is matched with a scientist who appears competent and communicative, the ability of expert witnesses to sway decisions is significant.

ACTIVITIES

1. Using the guest lecturer described in Activity 5, Chapter 13, test the person for expertise. Apply the standards described in this chapter.

2. Secure a transcript of a hearing with the help of your campus library. Note:
   - What kinds of backgrounds the experts who testified at the hearing had.
   - What surprised you about the hearing.
   - What kinds of oral testimony you found most convincing.
   - What effect (if any) the testimony had on the outcome of the issues being heard at this hearing.

3. Divide the class into groups of five for the purpose of role-playing. Each member of the group may choose a suitable expert background. Each group selects a topic area about which members will, with their assumed backgrounds, be expert enough to give testimony. Another group of five class members will act as a power board before whom the experts will testify. The board might be a senate subcommittee, a federal agency, and so forth. Each member of the expert group will present a four-to-five-minute statement to the power board. At the conclusion of each expert's statement, members of the board are free (and expected) to question the expert.
4. Secure a transcript of a court case where expert testimony was used to decide the verdict. Using the eight assumptions about testimony that our legal system makes, check to see what happened with the expert in the particular trial you examined.
5. Have your communication instructor arrange a field trip for the class to observe a courtroom trial where expert testimony is being heard. Note:
   - Whether the trial was as dramatic as you expected.
   - What sort of questions the attorneys from both sides used most frequently.
   - How the attorneys established a witness's expertise.
   - How the attorneys questioned a witness's expertise.
   - Whether you accepted the conclusions drawn by the expert.

NOTES

1. Francis L. Wellman, *The Art of Cross-Examination*, 4th ed. (New York: Collier Books, 1936), 94.

2. Richard D. Rieke and Malcolm O. Sillars, *Argumentation and the Decision Making Process*, 2nd ed. (Glenview, Ill.: Scott, Foresman, 1984), 237.

3. J. W. Patterson and David Zarefsky, *Contemporary Debate* (Boston: Houghton Mifflin, 1983), 35.

4. J. Vernon Jensen, *Argumentation* (New York: Van Nostrand, 1981), 95–96.

5. Kevin Tierney, *How To Be a Witness* (Dobbs Ferry, N.Y.: Oceana Publications, 1971), 33.

6. Austin J. Freeley, *Argumentation and Debate: Rational Decision Making*, 4th ed. (Belmont, Calif.: Wadsworth, 1976), 78.

7. Tierney, *How To Be a Witness*, 12.

8. Tierney, *How To Be A Witness*, 57.

9. Our discussion of agencies and their growing use of hearings to structure rules for policy change is based on Tierney, *How To Be a Witness*, 88–90.

10. We are indebted to Karen Kinard, Washington aide to Robert Edgar (D., Pa.), for her help in locating various committee hearing sources.

11. House Committee on Public Works and Transportation, Subcommittee on Investigations and Oversights, *Hearings*, 97th Cong., 1st sess. (Washington, D.C.: GPO, 1982).

12. House Committee on Veterans' Affairs, Subcommittee on Compensation, Pension and Insurance, *Hearings*, 98th Cong., 1st sess. (Washington, D.C.: GPO, 1983).

13. House Committee on Public Works and Transportation, Subcommittee on Investigations and Oversight, *Hearings*, 97th Cong., 1st sess. (Washington, D.C.: GPO, 1983).

14. Tierney, *How To Be a Witness*, 89.

15. Daniel Yarmey, *The Psychology of Eyewitness Testimony* (New York: Free Press, 1979), 179.

16. House Committee on Veterans' Affairs, Subcommittee on Compensation, Pension and Insurance, 98th Cong., 1st sess. (Washington, D.C.: GPO, 1983), 56.

Summary: Part III

Settings for Technical Communication

The technical professional communicates in a number of settings unique to the technical corporation. At the same time, such professionals are members of a larger community, and it too places demands on their energy and expertise. Four settings are identified as important environments where communicative competence must be displayed.

The first of these settings is the dyad, one-on-one communication. Like all humans, the technical person shares the need to establish relationships. Corporations are opportunities for at least three kinds of dyadic communications. Careers begin with the formal interview, which is—in part or wholly—conducted as a dyad. Interviewing involves a number of skills: preparation, handling of questions, and practicing expected courtesies.

Careers are maintained through individual dyads. These include interpersonal daily relationships, networking, and conflict management. Power structures of superior-subordinate relationships provide the setting for a professional career, and these relationships usually function in dyads. With the analysis of superior-subordinate responsibilities, we make five suggestions for coping with conflicts in power. Finally, the giving and taking of orders complete the dyadic power structure.

Group discussion is a very important and broad part of corporate business. We make suggestions for understanding what constitutes a group, what kinds there are, and what are the advantages and disadvantages of group communication.

Adaptations and new trends in modern corporate groups are important to a successful career. Quality circles, the bringing together of different levels of workers to share product-improvement ideas, seems a promising innovation. Various in-house and intercompany group meetings, usually taking place while a proposal or contract is in the works, reveal a form of hybrid public speech that corporations use. This format is widespread, in both the United States and Europe.

Participation and leadership in various group settings claim a great deal of the technical professional's time. We advocate preparation for problem solving by using the Dewey outline, preparing an agenda, and understanding parliamentary procedure.

Eventually, the technical professional is called upon to lead a group, as well as participate in it. The duties of both leader and participant are discussed, and we make suggestions to help competence develop.

Public communication is defined as more complex than the traditional picture of one person at a formal setting addressing a hushed audience. We see the public speech as a matter of assumptions made by participants. When one person takes on prime responsibility for message sending over an extended period of time, and the audience grants that person a hearing, the speech is public.

Public speeches take place in person, when the speaker and the audience are in one place, and in the mass media, when they are divided by distances. Both kinds are used widely in modern corporations.

The hybrid public speech occurs in group meetings. Fitting our definition, one person delivers an extended set of verbalizations; but, at some point, the audience expects to enter directly into the speech, since the audience members have final responsibility over the speech's consequence. This pattern of group/public speeches is probably the most frequent choice in the corporate setting.

The final setting for technical communicators is expert testimony. We define *expert* and give a number of tests to determine expertise. Technical persons are called upon to give opinions, supported by data and methodology, in both courts of law and at hearings. Some guidelines for preparing that testimony and delivering it are made.

So we come both to the end and to the beginning. For us, it is the end of these many pages of advice and theory; for you, it is the beginning or continuation of a technical career. We have told you without hesitation that we expect you to succeed. We have also told you that the issues discussed within these pages will, to a large degree, determine that success. Keep those issues in mind. We will look for you at professional meetings, honorary banquets, and on the pages of science newsletters. We will know you. You will be the one with extraordinary communication skills.

GLOSSARY

Each of the following words has multiple possibilities for definition. We present the sense in which we have used these terms in the text.

A

abstraction The process or operation of leaving out details. In language, the selection of symbols that increasingly make it difficult for a receiver to recognize what is being talked about.

abstraction ladder A pictorial model, formulated originally by Alfred Korzybski, showing the movement of language from low to high abstraction.

abstractor The creature, usually human, doing the observation and the selection.

action An observable behavior.

activation Setting into motion an action or thought process.

adaptors Nonverbals that help the person to cope with his or her self.

affective component A person's liking or emotional response to something.

affects Nonverbal behaviors that display a response to some inner state of feeling.

agenda A list of business items to be discussed in a group and the order in which they will be considered.

aggressor Usually a negative group role in which a person attacks another person verbally and increases group tension.

analogy A comparison between two similar things. In the reasoning process, a comparison of a known event to a similar event in order to demonstrate that the unknown is likely to function in the same way as the known.

anxiety A generalized reaction to the unknown.

assertiveness A basic set of behaviors that clarifies demands without diminishing the rights and feelings of another.

attitudes Enduring, generalized, learned predispositions toward some class of evaluative responses.

audience A composition of individual, separate, and variable perceivers of any communication.

authority The perceived power source. In the reasoning process, the power source exerts such influence that a receiver accepts a conclusion on the strength of the power source's opinion.

B

background Everything that has happened to the participants in a communication up to the present moment.

behavioral component The person's actions at any particular time.

blocker Usually a negative group role in which a person stops every approach to consensus.

body (of speech) The portion of an oral communication that comes between the beginning and the end.

brainstorming A non-evaluative group format that generates primary ideas by making lists of possible answers without discussing them further.

bylaws A document that spells out the process through which the organization and its members will operate.

C

case studies Personal stories that happen to someone else.

cause and effect A reasoning linkage that assumes that a given event is responsible for the occurrence of a second event.

channel The choice available to the sender through which the stimuli will be sent.

chronological organization A means of putting together the body of an oral communication so that items are arranged according to the time in which they happened.

clarifier Usually a positive group role in which a person says and does things that help group members sort out any ambiguities.

closed question An inquiry that requires a specific piece of information and restricts the expected response.

clown Usually a negative group role in which a person's humor is a serious distraction to group progress.

cognitive component A person's belief or knowledge about something.

cohesion In group process, the feeling of evolved belongingness to the group.

communication The sending and receiving of stimuli through some channel so that entropy is kept minimal.

communication process model A pictorial representation of how the Casagrandes believe communication flows.

conflict A recognized and communicated difference among persons, so that one or both of the dyadic members is conscious of opposition.

consensus In group process, when each person (although he or she might not completely like the decision) can live with it acceptably.

constants Factors that are always present when a given event happens.

consultants People hired for their expertise in order to give information and/or advice.

control room In business, a physical space used by project members to keep track of how each component of the project is progressing.

credibility The character and reputation a person is assumed to have.

cross examination Questioning in a law court by the opposition attorney.

customer Any purchaser of technical services or products.

customer review meeting A gathering of all those who have contracted with a customer for some services or products. The purchaser listens to reports on contract progress, and all participants make suggestions for improvements.

D

decoder Another name for the receiver of a communication.

deductive reasoning A reasoning function that is based on the linkage assumption that the general knowledge about a whole class of things makes it unnecessary to examine any single member of that class.

demographics The study of observably measurable factors in human populations.

direct examination In a court of law, the testimony presented initially under the direction of the friendly or allied attorney.

disorder In the physical world, the state of randomness into which things fall.

dyads Any combination of a single sender and a single receiver in any communication.

E

egocentrism One of the contemporary speaker goals, an aim that recognizes that every communicator monitors his or her own presentations and seeks to satisfy an inner feeling of worth with that presentation.

ego injury Conflict due to a perceived difference when one person assumes another does not value the original self.

emblems Nonverbal communications that can be translated directly into verbal words.

encoder Another name given to the sender of a communication.

energizer Usually a positive group role in which a person communicates something to the group that keeps the group going at an increased energy level.

entropy The tendency of physical events to fall into disorder. In communication, the disorder is the inability of stimuli to be decoded without loss.

evenhanded leadership A style of group direction that balances a concern for the solution reached with a concern for the feeling levels of the participants in that solution.

evidence The verifiable and empirical part of proof.

example One type of reasoning that asserts that a given event has happened once, and that this happening is demonstration of a conclusion.

expert A person whose opinions regarding some conclusion are trustworthy because of education, experience, or physical proximity.

extempore speech A type of oral communication delivery in which a speaker carefully prepares a set of ideas and documentation, but forms the actual wording at the time the speech is spoken.

external feedback One type of regulatory message sent back to the speaker from a source other than the speaker.

eye contact One of the nonverbal categories that communicates a potential for message in response to how long and in what direction eye gaze is regulated.

F

fact In popular language, an assertion that a given event actually happened as the reporter claims it did. In General Semantic theory, a statement that can be made only about the past or present by a person who made the perception directly and limits the language about that perception to what was actually perceived.

fallacy A distortion purposefully used in the reasoning process.

fear A specific anxiety reaction to a specific event.

feedback A response message sent back by the receiver to the original sender allowing the sender to monitor how the message sent was being decoded.

feedback, negative A set of response stimuli sent to the sender that discourages more of a similar message.

feedback, ordered A response message using language symbols.

feedback, positive A set of response stimuli sent to the sender that encourages more of a similar message.

feedback, random A response message using non-language symbols.

feedback, received That portion of the response stimuli taken in by the sender.

feedback, translated The response stimuli taken in by the sender and decoded into a response message.

formal meeting A grouping of people, often from different companies, meeting with knowledge about what is to be discussed, in what order the issues will be introduced, and what action is expected to result.

forum In a group setting, an open opportunity to speak for anybody who wishes to do so, even though that person is not a scheduled member of the presentation.

G

gatekeeper Usually a positive group role in which the person keeps track of where the group is in the agenda and says where they should go next.

general semantics A theoretical study of language formulated by mathematician/engineer Alfred Korzybski that asserts an inherent relationship between language, thought, and resultant behavior.

general to specific Another name for deductive reasoning.

goal What an individual or group of communicators seeks to accomplish as a result of the communication.

graph chart One type of widely used visual aid that presents data in a pictorial form depicting relationships.

group A distinct communication setting of three or more people, each able to communicate directly with the others, coming together in order to use each other for some purpose and, as a result of the communications, forming a sense of cohesion and group norms.

H

hands-off leadership A style of group direction that largely eliminates guidance on the part of the leader and places maximum concern for group-member feeling levels.

hearing A fact-finding meeting held for decision-makers to gather a range of opinions from various viewpoints.

heavy-handed leadership A style of group leadership that overtly regulates and directs a group. Maximum concern is usually for the quality of solution a group reaches at the expense of group feelings.

hidden version of me That portion of the self that neither the individual nor those around him or her realize is there.

high level of abstraction A level of language that gives very little indication of what the speaker is talking about. As a result, this language is likely to confuse and lead a listener into miscommunication.

human value Those things a person would choose as priority in comparison with other things.

I

illustrators Nonverbals, usually gestures, used along with verbal language to enforce or clarify the verbal.

impromptu speech One of the choices of delivery available to the speaker in which that speaker has made no preparation for the oral communication at hand.

inductive reasoning A type of reasoning that assumes that some observation of a specific event is necessary before any general conclusion about the whole class of events is possible.

inference In General Semantic theory, any statement about the past, present, or future, when the speaker has not made the perception about the thing being discussed, and does not regulate his/her language only to include that which was perceived.

information giver Usually a positive group role in which the person has important data, desired knowledge, and pertinent evidence, and is willing to supply it to other group members.

information seeker Usually a positive group role in which the person asks for needed knowledge, data, and evidence.

initiator Usually a positive group role in which the person introduces new ideas or information for group consideration.

in-person public communication That type of public communication in which the primary sender of the message and the receiver(s) are present in the same physical place.

in-studio taping That type of television video that takes place in a permanent videotaping facility.

internal feedback The reaction of a person to his or her own stimuli.

interpersonal communication The communication among dyads.

interpersonal relations The on-going associations built among dyads.

interview An in-person communication entered into by persons for some specific task goal and revolving around specific question and answer information.

intrapersonal communication Those communications that any person has with the self.

introducer The person who presents the next speaker to the audience.

J

jargon The highly specialized terminology evolved by some disciplines and usually unfamiliar to those outside that discipline.

K

kinesics One of the categories of nonverbal communication by which messages can be decoded by observation of muscle movements.

L

language Any stimulus to which a community has agreed to give meaning and rule-governed structure.

leadership A person's willingness to risk a communication that has an impact and results in influencing the action of another.

listening The translation of a message function that a receiver performs.

logos According to Aristotle, proof that appeals to the thinking and reasoning part of humans.

low level of abstraction That kind of verbal language that names and gives clear observational descriptions of events being talked about.

M

main motion In parliamentary procedure, a low-precedent verbal statement that details some business that the mover of the motion wants the group to consider.

manuscript speech One of the choices of delivery in which the entire speech is written out ahead of time and read to the audience.

mass media That kind of public communication in which the sender and the receiver(s) are physically separated by distance and so use written or machine channels for the transmission of the stimuli.

memorized speech One of the choices of delivery in which a manuscript is committed to memory and delivered without any notes.

message received That portion of the communication model in which the receiver translates meaning for the stimuli sent.

message sent That portion of the communication model that represents what the sender intends the stimuli to mean.

methodology The plan of action and processes used to complete a given task.

minutes The written record of a meeting.

monopolizer Usually a negative group role in which a person uses most of the meeting time without giving other group members a turn to speak.

Monroe's Motivated Sequence An organizational pattern for persuasive speeches developed by Alan Monroe.

motivation In the contemporary speaker goal theory, the changes a sender aims to make in the thinking and knowledge of the receiver.

N

need A state of unbalance in any system which, if left unbalanced, will frustrate the system in its milder form and shut down the system in its most severe form.

negotiation The verbal process of giving up something in order to demand something else.

network A communication pattern that uses personal dyads in series to transmit information.

noise In traditional communication theory, anything that contributes to the distortion of a message.

nonparticipant Usually a negative group role in which a person says little or nothing during a meeting, thereby denying the group the benefit of the person's ideas.

nonverbal communication A communication in which the stimuli, random or ordered, are unrelated to the content of words.

norms Informal rules of procedure developed by group members for the operation of their group meeting.

O

objects In nonverbal communication, the messages decoded through the use of physical things. In visual aids, those physical things used by a speaker to demonstrate and clarify what is being discussed.

open-ended question A type of inquiry that permits a range of unrestricted responses.

opinion A judgment made by someone concerning something. In proof theory, one of the two segments of evidence.

on-site taping That type of television video prepared at the place a given event is happening rather than at a permanent studio.

order A verbalization that specifies what a superior wants a subordinate to do.

ordered stimuli Bits of activating energy to which communities agree to give meaning and rule-governed structure.

P

panel A discussion among experts on a topic, usually for the benefit of an audience. Each expert, however, can contribute to the discussion whenever he or she wishes.

paper/flip charts Informal visual aids made of paper and used to present materials without the use of machines.

paralanguage One of the categories of nonverbal communication in which message translation is possible through voice overtones such as pitch, pronunciation, rate, noises between words, and volume.

paraphrase An effective listening device to tell the speaker (sender) what you thought was being said.

parliamentary procedure A set of rules and procedures by which an organization can regulate the flow of business during a meeting.

pathos According to Aristotle, a kind of proof elicited by a speaker through the raising of listener emotion.

peacemaker Usually a positive group role in which the person acts to mediate conflicts within the group.

persuasion One of the traditional speaker goals in which some specific action is sought from the receiver(s). In the contemporary speaker goal theory, this goal is called activation.

playperson Usually a negative group role in which a person wishes to play with the process of the group, thereby keeping the group from any movement toward consensus.

power structures A factor in every dyad, distinguishing the one who influences the thought and action of the other from the one being influenced.

prime message sender In a public communication, that person who has the major responsibility for acting as sender for some protracted length of time.

problem to solution A means of organizing the body of the speech or a written report so an analysis of some difficulty is discussed as a preparation for a suggested solution to that difficulty.

process A state of continual change.

pro-con development A means of organizing the body of a speech or written report that presents both sides of a given problem and compares them before discussing a suggested solution.

productivity An individual group member's evaluation of satisfaction with the content and feeling levels of any group.

proof The evidence and reasoning portions of a communication.

proxemics A nonverbal category that notes the possibility for message decoding based on the manipulation of space.

public communication Any communication in which one person has the prime responsibility to act as the source of oral messages for an extended period and the receivers grant the person that right.

Q

quality circles (quality control circles) A type of group meeting that brings together all different levels of business people who are involved in the production of some product, in order to hear how the product can be improved from those who actually work to produce it.

R

random stimuli Bits of activating energy that occur without regularity or structural agreement.

reasoning The process of linking known information in order to speculate on the unknown.

receiver The one who accidentally or intentionally is the recipient of the stimulus.

recognition seeker Usually a negative group role in which a person uses the group to increase his or her ego.

record keeper Usually a positive group role in which a person assumes the responsibility for maintaining an accurate recording of the group's activities.

reflective thought pattern A speculation made by John Dewey on how people think during problem solving.

regulators Nonverbal behaviors that signal turn-taking during conversations.

report form A version of the problem-to-solution organizational pattern used frequently in the corporate world.

review meeting A type of business group that assembles all those involved in some common project in order to check the progress of their various tasks regarding that project.

rhetoric The name traditionally given to persuasion.

S

sanctions Use of threats or punishment.

scientific method A system of investigation that requires a carefully worded statement of some assumption, the testing of that assumption through observational and reproducible procedures, and the adjustment of the statement based on the outcome of the testing.

self-concept The picture that one has of him- or herself.

sender In communication theory, the originator, the beginning point, the initiator, the one who starts the communication on its way.

setting The environment in which the communication takes place.

sign A natural and relatively random happening that holds communicative potential.

signs, nonreciprocal An occurrence of two signs when one variable is a reliable sign of the other but the second is not.

signs, reciprocal An occurrence when the two variables work so that both are reliable signs of the other.

skip level meetings A type of business group meeting that brings together people at some level with the boss of that group's supervisor. Since the supervisor is absent, this format gives the subordinates a chance to express whatever they wish to their boss's boss.

speech anxiety A fear of oral presentation, especially public communication.

stimuli Bits of energy capable of activating a response.

subordinate The person influenced to act by the superior.

superior The person having the influence to affect another's actions.

supporter Usually a positive group role in which a person allies with another person or set of viewpoints to sort out where the strength of support is.

symbol An event that is taken by humans to represent something else.

symposium A discussion among experts on a topic, usually for the benefit of an audience, in which each expert speaks on some preplanned portion of the topic without interruption.

system theory An assumption that every part of some entity affects and is affected by all the other parts.

T

task meeting A type of business group in which all people involved in some common specialized activity assemble in order to review their progress toward completion of that activity.

telecommunication A system consisting of telephone networks through which oral messages, hard copy messages, graphics, computer files, and telegrams are sent.

tension reliever Usually a positive group role in which a person communicates something that allows the group to return to a lesser state of stress.

testimony A term used to designate the statements a witness makes about the specific area of his or her special knowledge.

translation of the stimulus That portion of the communication model in which the receiver decodes a message from the stimulus.

transparency An extremely common visual aid for technical presentations, made on a clear sheet for use with an overhead projector for the purpose of enhancing a speaker's message.

U

unit meetings In business groups, the narrowest and most specialized assemblies of people brought together to check the progress of a given unit of work that all members perform.

unprepared one Usually a negative group role in which a person has not done prior thinking and accumulation of information.

V

value A human choosing system that selects certain things as priorities over others.

variable A specific component.

variable, constant A specific component always present in some event.

variable, transient A specific component that might or might not be present in a given event.

verbal communication Those communications involving words.

Version of Me model A pictorial representation of the individual self according to Casagrande thinking.

visual aid A physical object, film, chart, transparency, picture, videotape, or model used to enhance the speaker's presentation.

W

words A particular unit of sounds or graphics that forms the smallest agreed-upon unit of meaning through which a concept is expressed in verbal language.

INDEX

abstraction, 65–69
  ladder, 66–68
abstractor, 65–66
acceptance of contract, 139–141
acceptance of ideas, 136–138
acceptance of image, 134–136
acceptance of methodology, 138–139
action as a speaker goal, 130–142
  in discussion questions, 250–251
Adams, Douglas, 79
adaptors, 49
affects, 49
agenda, 244, 264, 266–272, 273
  Dewey-based, 268–272
  hidden, 268
  informal, 268
  parliamentary process, 272–278
  problem-solving meeting, 244
  types of, 268–278
analogy as part of reasoning, 174, 180
anxiety in speaking, 109–113
Aristotle, 129, 164–165, 295
artistic and nonartistic proofs, 164–165
audience, 120–142
  analysis, 121–128
  attitudes, 123–124
  demographics, 122–123
  response, 128–129
audio aids, 184–185
authority as expert, 318–337
  in reasoning, 174–175, 181

background, 21–22
bias in hearings, 334–336
body of speech, 153–158
body (human) types, 52
brainstorming, 242
Brooks, William D., 315
Bruskin Report, The, 109
bylaws, 272–273, 274

career development in technology, 3–5
  and interpersonal relations, 219–226
  and networks, 220–221
Carlson, Bruce, 29
case studies as proof, 174, 180
cause-and-effect reasoning, 172–173, 179
cautions in verbal and nonverbal messages, 61–62
channels of communication, 19–20, 23–24, 184
closed question, 216–217
cohesion in groups, 240–241
communication
  aids, 184–206
  definition, 40
  group, 9
  helical, 33–34
  interpersonal, 8–9
  intrapersonal, 7–8
  losses, 221
  to meet professional expectations, 11–14
  nonverbal, 48–62
  as part of life, 6–11
  public, 9–11, 294–315
  verbal, 62–75
communication process model, 33, 40–42
communication variables, 19–40
  constant, 19–34
  focused, 296–298
  transient, 34–40
conflict management, 221–226
  ego injury, 223–224
  interpretation differences, 222–223
  methodology, 223
  personal values, 224–225
  physical impossibilities, 222
consensus in groups, 241
consultants, 150–151
contract as goal of professional communication, 139–141
control room, 255–256
corporate forms of group meetings, 251–263
  cooperative contract, 259
  crisis, 252–253
  customer review, 259–260
  international, 260–263
  orientation, 253
  progress, 253–256
  quality circles, 257–258
  regularly scheduled, 252
credibility
  as acceptance of corporate image, 134–135
  as acceptance of personal image, 135–136
  as education, 320
  and expertise, 321–323, 326, 329
  in interviews, 218
  as a professional expectation, 12–14
  in public speaking, 298, 304
crisis meetings, 252–253
cross-examination, 324–325, 326–327
crowds (as opposed to groups), 239–241
customer review meeting, 259–260

decoder, 24–26
deductive reasoning, 171–172
definitions of terms in glossary, 342–350
Deming, W. Edwards, 257–258
demographics, 122–123
Dewey-based agenda, 268–272
dialect as paralanguage, 50–52
direct examination, 326
discussion questions, 248–251
  fact, 249
  policy, 250–251
  value, 249–250
Disney, Walt, 185–190
DuBrin, Andrew, 226–227
dyads, 213–235
  career development, 213–221
  and conflict, 221–226
  as interpersonal communication, 219–221
  interviewing, 213–219
  power structures, 226–235

education as expertise factor, 320
effective aids, 185–190

egocentrism as speaker goal, 130
ego injury in conflicts, 223–224
Einstein, Albert, 65–66, 69–70
emblems, 48
encoder, 20–21
entropy, 29–31, 51, 52, 83–85, 151–152, 234–235, 255–256, 335
  control of, 126–128, 172, 284
  sources of, 88, 104, 298
entropic noise, 29–31
ethos, 164
evidence, 161–170
  fact, 165–166
  opinion, 166–167
  as part of proof, 165–170
  strong/weak, 167–170
example as part of reasoning, 174, 180
expediency and professional expectations, 11–12
experience as expertise factor, 320
experiential effect, 34–35
expert, 318–337
  competence sources, 319–321
  in hearings, 332–336
  in law courts, 324–332
  testimony, 324–336
  testing for expertise, 321–324
extempore speech, 300–301
external feedback, 83
eye contact, 56–58
eyewitness, 321, 322, 324

Fabun, Don, 36
fact and inference, 69–75
  as discussion questions, 249
  as evidence, 165–166
fallacy, 175–181
fear, 110–116
  coping mechanisms, 110
feedback, 26–29, 81–83, 88–89
  external, 83
  internal, 82–83
  negative, 28
  ordered, 26
  positive, 28
  random, 26
  received, 81
  sent, 81
Fisher, B. Aubrey, 241
formal meeting, 266–278
forum, 243
Freeley, Austin J., 327

General Electric, 197, 199, 200, 251, 256
General Motors, 310
general semantics, 63–69
gestures, 60–61
Gibb, Jack R., 111
Gilbreth, Lilian M., 13
goals in communication, 128–141
  contemporary, 130–134, 295
  professional, 134–141
  traditional, 129–130, 295
graph chart, 195–199
groups, 9, 238–293
  advantages/disadvantages of, 244–248
  components of, 239–242
  consensus, 241
  corporate forms of, 251–263
  discussion questions, 248–251
  international, 257–258, 260–263
  leadership, 281–289
  member participation, 289–291
  parliamentary procedure, 272–278
  preparation for, 266–272
  productivity, 268, 289
  roles, 279–281
  types of, 242–244

hearing (as opposed to listening), 81
hearings, 318–337
  experts in, 332–336
helix, 32–33
high level of abstraction, 65–69
human values and needs, 104–108
Hunt, Gary T., 295
hybrid public speech, 254, 259, 260, 306–307

illustrators, 49
image as credibility, 134–141
  acceptance, 134
  acceptance of ideas, 136–138
  acceptance of contract, 139–141
  acceptance of methodology, 138–139
  corporate, 134–135
  personal, 135–136
impromptu speech, 301–302
inductive reasoning, 171
inevitability of change, 63–65
inference and fact, 69–75
influence and power, 226–227
information and traditional speaker goal, 129–130
  and public audience speeches, 308–309
in-studio TV taping, 311–313
intercultural differences in business meetings, 260–263
internal feedback, 82–83
interpersonal communication, 213–235
interviews as a professional beginning, 213–219
  courtesies, 218–219
  preparation for, 214–216
  questions and answers, 216–218
intrapersonal communication, 7–8, 99–117
introducer, 304
Ishikawa, Kaoru, 258

jargon, 68–69, 298, 332
Jenson, J. Vernon, 326
Johnson, Wendell, 7–8, 81–82, 91

Keltner, John, 290–291
kinesics, 60–61
Korzybski, Alfred, 63–66
Kuhn, Thomas, 136–138

Labov, William, 69
language, 23, 47, 64, 65–69
  nonverbal, 23, 48–61
  verbal, 23, 62–75
law courts, 318–337
  experts in, 324–332
leadership in groups, 281–289
  assigned, 282–284
  emerged, 284–286
  responsibility, 282–284
leadership styles, 286–289
  even-handed, 287
  hands off, 286–287
  heavy-handed, 286
  and personality, 288–289
levels of abstraction, 65–69
listening, 79–91
  effectively, 84–91
  to others, 83
  to yourself, 82–83

litigation, 324
logos, 165
low level of abstraction, 65–69

main motion, 274–278
manuscript speech, 299
Maslow's need theory, 106–108
mass media, 10–11, 310–313
May, Rollo, 213–214
McCrosky, James, 113–114
meetings
  crisis, 252–253
  customer review, 259–260
  formal, 266–278
  in-house, 251–258
  inter-company, 258–263
  international variations, 260–263
  orientation, 253
  progress, 253–256
  quality circles, 257–258
  regularly scheduled, 252
  skip-level, 256–257
  staff, 254–255
  task leader, 254
  unit, 255–256
meeting agenda, 244, 268–278
memorized speech, 300
message received, 26–29
message sent, 22
microphone use, 304–305
minutes, 273–274
model making, 17–19
  Casagrande communication process, 17–42
  Shannon-Weaver, 18
  versions of me, 100–104
Monroe's motivated sequence, 156–158
motivation as a speaker goal, 130–131
Myers-Briggs Indicator, 288–289

needs, human, 106–108
negotiation, 290–291
networks, 220–221
Nichols, Ralph, 79
noise and entropy, 29–31
nonverbal communication, 48–62
  adaptors, 49
  affects, 49
  emblems, 48
  illustrators, 49
  in interviewing, 215, 218–219
  in public speeches, 297–298
  regulators, 49
norms in groups, 240

objects
  as nonverbal communication, 52–55
  as visual aids, 201–202
one-to-one encounters, 8–9, 52–55
on-site television taping, 311
open-ended question, 217
opinion as part of proof, 166–167
  in testimony, 325
oral communication, 12
oral report form, 155–156
ordered stimuli, 23
order giving and taking, 232–235
organization of topics, 151–158
  beginnings and endings, 152
  chronological, 153
  general to specific, 154
  motivational pattern, 156–158
  oral report form, 155–156
  pro-con development, 153
  spatial, 153
  written report form, 154–155
overt checking in listening, 90–91

panel, 243
paper/flip charts, 202
paralanguage, 50–52
paraphrase in listening, 90
parliamentary procedure, 272–278
  adopting motions, 277–278
  agenda, 273
  blocking motions, 277
  making a motion, 274–276
  minutes, 273–274
  quorum, 278
  reports, 274
  safeguarding rights, 276–277
  summary of common motions, 275
pathos, 164–165
Patterson, J. W., 325
perceptual bias, 35
persuasion, 129–142
  as contemporary goal, 130
  and credibility, 134–142
  in hearings, 335
  in hybrid public speeches, 307, 314
  in job interviews, 214–219
  as part of all communication, 129–130
  in public speeches, 295–315
  as traditional goal, 129
physical position as expertise factor, 321
policy discussion questions, 250–251
  formation and experts, 321
power, 226–235
  sources of, 227
  superior/subordinate interaction, 227–231
  giving and taking orders, 232–235
presentational charts, 190–201
  bar, 194
  line, 196–199
  pie, 195
process nature of the world, 63–65
productivity in groups, 268, 289
proof, 161–181
  components of, 163–164
  sources of, 164–165
proposal in contract, 139–141, 253
  as sales campaign, 305–306
proxemics, 58–60
public communication, 10, 295–315
  formalized goal, 296–297
  and general public audiences, 308–310
  hybrid speech, 306–307
  and microphones, 304–305
  nonverbal messages, 297–298
  and oratory, 295, 301
  and professional organizations, 303–305
  and sales contract opportunities, 305–308
  specific audience, 297
  and television, 310–313
  types of delivery, 299–302
  verbalized messages, 298
  visual and audio aids, 302
public hybrid speech, 254, 259, 260, 306–307

quality circles (quality control circles), 257–258, 320
questions in interviews, 216–218
  closed, 217
  in law courts and hearings, 324–337
  loaded, 331
  open-ended, 217–218
quiet people, 113–117

random sound perceptions, 6–7, 45–47
random stimuli, 22, 40, 45
Rankin, Paul, 5–6, 79
reasoning as proof, 170–181
  categories of inductive and deductive, 171–172
  fallacy, 175–178
  testing for strength of linkage, 178–181
  types, 172–175
receiver, 24–26
reflective thought pattern (Dewey), 268–272
Rega, Paul, 238
regulators, 49
research and organization of topics, 147–158
resources for research, 147–151
  books, 148
  consultants, 150–151
  government sponsored research, 149
  interview, 149–150
  journals, 148
  overlooked, 149
  personal experiences, 150
rhetoric, 129, 295
Rieke, Richard D., 324–325
roles in groups, 279–281
  positive, 280
  negative, 280–281
Rosenfeld, Lawrence, 279
Ross, Raymond, 80, 310

Scheidel, Thomas, 105
scientific method, 163–164, 271
self-talk, 7–8, 26–29
sender, 20–21, 81
settings for technical communication, 213–337
  dyads, 213–235
  groups, 238–292
  public communication, 294–315
  expert testimony, 318–336
Shannon-Weaver model, 18
Sheldon, William, 52
shyness and quietness, 113–117
signs and symbols, 45–48
signs as part of reasoning, 172, 179
Sillars, Malcolm, 324–325
skip-level meetings, 256–257
Socrates, 99
speaker goals, 128–141
  activation, 131–132
  ego satisfaction, 130
  immediate and eventual, 132–134
  motivation, 130–131
  professional, 134–141
  traditional, 129–130
speaker bureaus as public relations resources, 309
speech anxiety, 109–113
speech delivery choices, 299–302
  manuscript, 299
  memorized, 300
  extempore, 300–301
  impromptu, 301–302
spontaneous group meetings, 242–243
Stevens, Leonard, 79
stimuli (defined), 22
  random and ordered as language, 22–23, 45–46
styles of leadership, 286–289
  even-handed, 287–288
  hands-off, 286–287
  heavy-handed, 286
  and personality, 288–289
subcategories of nonverbal communication, 49–61
subordinate/superior interaction, 227–235
subpoena, 334
symbols and language, 47–48
symposium, 243
system theory, 45

task meeting, 243–244
telecommunications, 204–206
television speaking, 310–315
  on-site taping, 311
  in-studio taping, 311–313
testimonial, 319–321
testimony, 324–336
  expert, 318–337
testing for expertise, 321–323
Tierney, Kevin, 326–327, 329, 331, 334
topics, 145–147
  mandated, 145–146
  organization, 151–158
  personal choice, 146–147
  research, 147–151
touch, 55–56
transient variables, 34–40
  experiential effect, 34–35
  perceptual bias, 35, 38
  professional expectations, 38–40
translation of the stimulus, 81
transparency, 190–201
Trautmann, Fredrick, 296

unit meetings, 255–256
unprepared group member, 281, 291

values, 104–106, 224–225
  as discussion questions, 249–250
  and ethics, 106, 108
variables
  constant, 19–34
  focused, 296–298
  transient, 34–40
verbal communication, 62–75
Verderber, Kathleen, 222, 225–226
Verderber, Rudolph, 222, 225–226
versions of me, 100–103
visual aids, 184–206
  flip chart, 202
  model/object, 201–202
  photograph/film, 202–204
  transparency, 190–201
  word chart, 190–195

Watson, Thomas, 290
Weinberg, Harry, 74
Wellman, Francis, 319
Williams, Frederick, 310
witness as expert, 318–337
  preparation for testimony, 328–332
written report form, 154–155

Yarmey, Daniel, 335

Zannes, Estelle, 115–116
Zarefsky, David, 325